W9-BYD-778

Insight into Eyesight

Insight into Eyesight

The Patient's Guide to Visual Disorders

Paul E. Michelson, M.D.

Nelson-Hall nh Chicago

Library of Congress Cataloging in Publication Data

Michelson, Paul E
Insight into eyesight.

Includes index.
1. Vision disorders. I. Title.
RE91.M52 617.7 79-22436
ISBN 0-88229-567-5

Manufactured in the United States of America

10 9 8 7 6 5 4 3 2 1

To Shelley, David and Jonathan, Rose and Harry for their support and indulgence, and to my patients for their inspiration and encouragement.

Contents

Introduction

This book is part of a growing trend in medicine to inform patients and the general public about disease and treatment principles. In medicine, as in other fields, there has been a technological explosion: we can cure more diseases; we can diagnose them more accurately; we can control and prolong function longer in incurable diseases. But we still have a long way to go. The publicity attending some of the advances and the illusions about modern medicine that are fostered by the entertainment field have created a credibility gap. Some people have come to believe that they can be completely cured of any disease or injury. It is devastating to harbor these unrealistic expectations or to disregard complications that must be risked if a cure is attempted. The exciting new advances that have been made in combating disease are not without risks and side effects, the usual price of progress. After all, virtually no treatment known to man is universally applicable without some risk.

It is the conviction of this author and many of his colleagues that a communication gap rather than a credibility gap cripples the relationship between some patients and the medical profession. To the extent that patients do not understand or acknowledge their illnesses, they cannot be expected to comply with demanding treatment regimens. And to the extent that they do not understand or acknowledge all of the implications and risks involved in certain operations or therapies, they are not capable of participating in crucial decisions. Patients' enlightened participation in the decision-making process is especially important in the many areas of medicine that cannot be simplified to one obvious and imperative therapeutic approach.

It is impossible to translate all medical knowledge succinctly into lay terms. Even the specialty of ophthalmology cannot be condensed into volumes of texts, since some information is outdated by the time it gets into print. Specialized periodicals and journals, general medical periodicals, courses, lectures, and meetings are all part of the continuing education of an up-to-date eye physician. This book is not written to help patients second-guess their eye doctor or prescribe their own treatment. Its purpose is to educate them so that they may more readily understand and accept their treatment plan and participate in any decisions about it. Patient input should render the treatment more effective and appropriate.

Another issue of concern to the medical profession and the lay public alike is the rapidly escalating cost of medical care. It is sheer folly and a disservice to society to pick scapegoats for this dilemma. By and large, the mounting costs are due not to the greed and profiteering of physicians, hospitals, pharmacists, nursing home operators, or drug companies, but to inflation and improved care. Great technological breakthroughs have been made in medicine as a whole, but it is expensive to develop and implement new treatment modalities. A patient who dies outright from cancer or heart attack incurs little expense; one whose life is saved or extended through intricate regimens of special drug therapy, surgery, irradiation, and intensive care incurs a staggering bill. An individual with advanced diabetic retinopathy went blind for the cost of only a few office visits no more than a decade ago. Today the potential for saving sight exists, but it involves continuous care, sophisticated evaluation, laser therapy, and possibly major retinal or vitreous surgery. This progress is responsible for increased medical costs that society bears.

This book is meant to convey the excitement and hope that have permeated ophthalmology, only one small segment of the medical profession, since new advances have enabled us to reduce blindness and eye disease. Once we appreciate what we are getting for our medical dollars, we may then be better prepared to decide what we can afford and how our taxes should be spent.

The scope of eye disease and visual disorders in the United States alone is staggering in both human and economic terms. Next to cancer, blindness is the most feared affliction. Approximately one and a half million Americans are functionally blind, and another ten million have significant uncorrectable impairments of vision. Several years ago there were about two million acute eye conditions, two million chronic visual problems, and two million eye injuries recorded in the United States. There are some seventy-five million eye examinations per year performed in the United States by ophthalmologists and optometrists for conditions varying from simple refractive errors or routine checkups to surgical eye disease. The direct cost of this eye care five years ago amounted to several billion dollars; the indirect cost of lost wages, disability, governmental support for visually impaired individuals, and related factors was estimated to amount to another billion-plus.

The author wishes to acknowledge the incalculable help of his chief proofreader and editor, Shelley Michelson; his indulgent illustrator, Sylvia Adler; and understanding typists, Ms. Margaret Bennett, Tana McCaffery, and Aleica Barber.

1
Definitions

Even highly educated laypeople often confuse the titles and functions of the various professionals and paraprofessionals involved in vision care. The following definitions will distinguish the types of practitioners.

Ophthalmologists, also called oculists, are medical doctors specializing in eye care. They are the only professionals who can perform a complete medical eye exam (including the prescription and fitting of spectacles and contact lenses), diagnose and treat all visual disorders, medical and surgical diseases, and injuries of the eye. As physicians, ophthalmologists note any disease process or drug therapy that may threaten vision. Life-threatening diseases such as high blood pressure, diabetes, or brain tumors are often detected in the course of a complete eye examination, as are the effects of certain medications.

An ophthalmologist's professional education consists of a suitable premedical undergraduate education leading to a bachelor's degree, followed by at least four years of medical school leading to an M.D. degree. After medical school, a year of internship usually precedes a minimum of three years of residency training devoted solely to eye diseases and surgery.

Upon completing this training, an ophthalmologist is considered board-eligible and may then take comprehensive written and oral examinations leading to board certification, an acknowledgment that he has the training, knowledge, and ability to practice ophthalmology. Some very fine ophthalmologists never bothered to take their board exams, but today most young ophthalmologists wish to establish their qualifications by being board-certified. A directory of board-certified ophthalmologists is available through the American Board of Ophthalmology. Once board-certified, an ophthalmologist may enhance his credentials and affirm his commitment to quality practice and continuing education by applying for honorary fellowship in the American Academy of Ophthalmology, the American College of Surgeons, and other organizations. A patient newly arrived in a community will find such qualifications helpful in selecting a doctor. Ophthalmologists need not be board-certified to practice, and some qualified practitioners have simply never bothered. A personal recommendation from a satisfied acquaintance or another physician in the community is also a suitable way to find an eye specialist.

Optometrists, doctors of optometry, are trained to detect visual problems and to diagnose and treat refractive errors and perceptual dysfunction. Optometrists must spend two to four years in an undergraduate preoptometric program and four years in a school of optometry. They are licensed to prescribe spectacles, contact lenses, and low-vision aids and to use visual training, orthoptic exercises, and other nonmedical modalities in the treatment of some forms of strabismus (crossed eyes) and reading disorders. In some states, optometrists are licensed to use topical ocular drugs. Many ocular and systemic disorders requiring referral to an ophthalmologist or other medical practitioner can be detected in a thorough optometric examination.

Opticians are also known as ophthalmic dispensers or dispensing opticians. They are trained to fill prescriptions for spectacles and to fit contact lenses and frames. These paraprofessionals can help the individual choose appropriate frame and lens design, verify the specifications of the final product, and make adjustments or repairs.

2 The Ophthalmologic Examination

The Eye Examination

The eye examination is quite different from most tests patients are accustomed to. Their cooperation is needed to determine the need for glasses, to detect defects in the visual field, and to satisfy all other elements of a full evaluation. Because many people are so sensitive about their eyes and any manipulation of even the eyelids, patients should be assured that no part of the examination is harmful or painful. Familiarity with the instruments and procedures involved in eye evaluation will relieve anxiety and make the experience interesting if not enjoyable. A comprehensive eye examination can yield vital information about one's general as well as visual health. It should not be sold short as a mere checkup for glasses.

The History

A patient's history is of obvious importance to the doctor. The first order of business is to find out what has brought the patient in for an exam. Even if the visit is prompted by nothing more urgent than a simple desire for a routine checkup and reassurance that no problems are developing insidiously, the doctor will want an elaboration of any symptoms being experienced—when they started, whether they developed suddenly or progressively, and related questions. Once the full story of the present illness has been obtained, other general and sometimes more important information must be elicited. A family history will reveal whether any blood relatives have had genetic diseases that the patient might develop or transmit to his offspring. Such eye problems as true night blindness (*retinitis pigmentosa*), juvenile cataracts, early macular and retinal degenerations, certain congenital eye tumors, and simple refractive errors have definite hereditary patterns. Other problems such as glaucoma usually have no common pattern of heredity but may be statistically more likely given a positive family history.

A general medical profile is also illuminating. Many diseases affect the eyes (see chapter 11), and many medications used for seemingly unrelated problems may have serious ocular side effects. A patient should tell the eye doctor about all medications used regularly and about all present or past general illnesses. Years of training help the doctor recognize significant features of the patient's history.

Measurement of Vision

The level of vision is most commonly expressed as 20/20 or a similar figure, but in fact this figure represents only part of the vision. This fractional notation refers only to visual *acuity*, the ability to discern fine detail. An equally important part of vision is visual *field,* the whole panorama through which we see. Only the central several degrees of each eye's normal 160 to 170-degree horizontal by 120-degree vertical expanse of vision has good acuity. The rest of the field serves as a guide, warning, and orientation system even though it has poor acuity and can distinguish no fine detail. The best visual acuity will still result in legal and functional blindness if visual field loss is great enough. Similarly, loss of central vision and acuity, which results in an inability to read or see detail, also constitutes definite disability. We must assess both acuity and field, then, in the measurement of vision.

Visual Acuity Measurements in Adults and Children

Visual acuity is expressed as a fraction: the denominator, or lower number, indicates the distance from which the normal eye can discern a given figure, and the upper number is the distance at which the patient is able to see it. Thus, 20/20 means that at twenty feet (or miles, or any unit) a normal eye can see the objects in question, and so can the patient. 20/40 indicates impaired vision in that the patient at twenty feet sees only as much as the normal eye can see from forty feet. 20/200 would be still worse, as the patient now sees at twenty feet what a normal eye can see from as far away as two hundred feet. If the standardized figures cannot be seen at twenty feet, they can be brought closer. Thus, 3/200 indicates that the patient sees at three feet what a normal eye can see at two hundred feet. Once vision is this poor, it may be expressed as the ability to count the fingers held up at a given distance. In worse cases, it is gauged by the ability to detect motion or light from darkness.

Each of the letters in the standardized eye chart is so constructed that the width of each line and the "gaps," or separations, in the letter measure a given number of degrees. The common English letter chart is called the Snellen chart, after its originator. Similar charts exist in other languages.

Several charts have been devised to test visual acuity in children and illiterates. The most common is the E chart, which presents different-sized E's that are oriented variably in the four directions. Subjects are taught to hold their hand with the fingers pointing in the same direction as the three bars of the letter E being presented to them (figure 2–1). Children who cannot learn or will not cooperate often respond to picture cards or charts that have figures calibrated for the width of lines and separations to represent the same acuity measurements as the standardized letters.

No good subjective measurement of vision is possible with very small or uncooperative children, but a trained observer can at least assess the comparable level of vision in each eye by watching the child follow objects with one eye and then with both. A child with markedly poor vision in one eye, from amblyopia for example (see chapter 7), will resist having the better eye covered. He may not exhibit normal, steady fixation or react to a threatening gesture directed toward the eye. Special rotating drums called optokinetic drums are used to elicit the normal quick rhythmic jumps of the eyes that occur when a staccato sequence of lines or images passes across the field of view. Normally, we can notice this movement

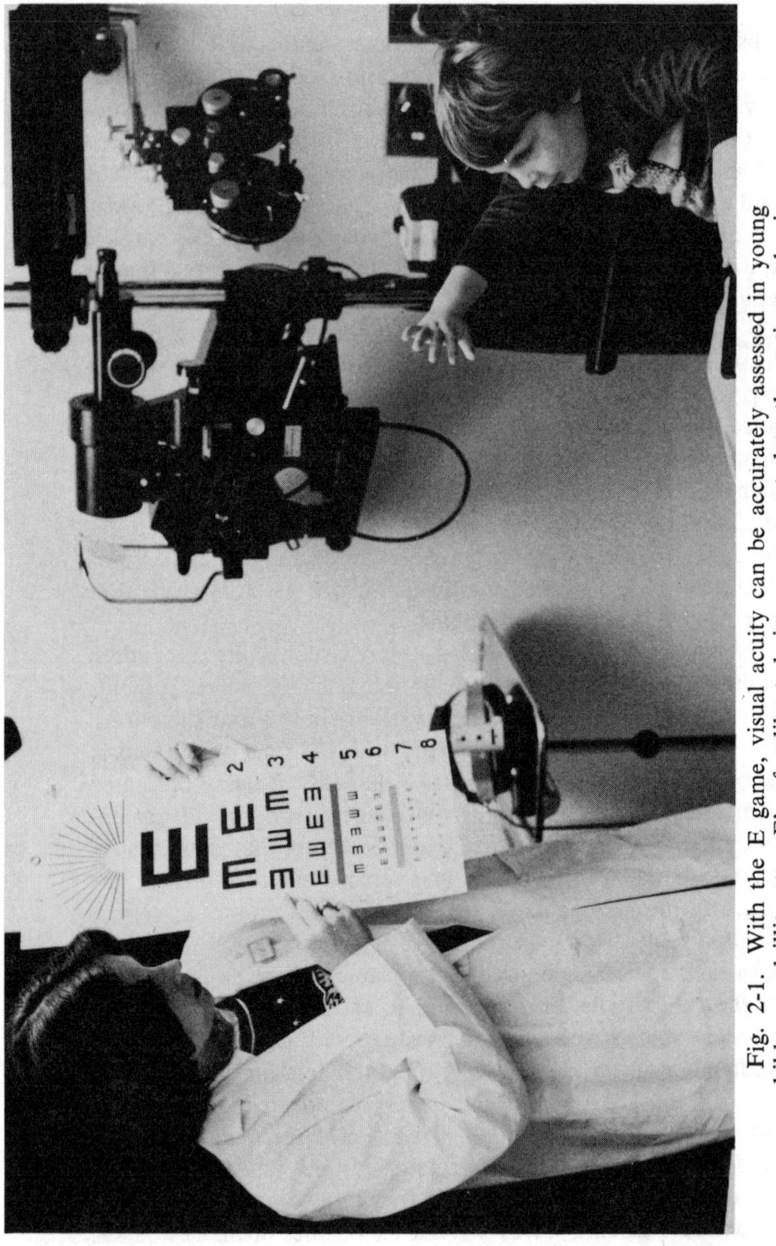

Fig. 2-1. With the E game, visual acuity can be accurately assessed in young children and illiterates. E's of calibrated size are presented to the patient, who is taught to respond by pointing his hand in the same direction as the three bars of the letter.

in the eyes of an individual who is watching some relatively close objects such as utility poles from the window of a moving vehicle. If the normal optokinetic response is absent, some visual problem may exist.

In acuity measurements 20 is used because twenty feet is a practical distance at which to measure distance acuity. Theoretically, no focusing should be necessary for a true distance acuity. Near focusing occurs because the light rays reflected from a given object are still diverging when they hit the eye. Once an object gets eighteen to twenty feet away, the rays from it that enter the eye are nearly parallel—like those from a more distant object (see chapter 12). The notation used by convention in Europe is six meters, which is about the same as twenty feet. Normal vision there is expressed as 6/6. This vision could also be expressed in decimal form, where 1.0 equals 6/6 or 20/20; 0.5, then, is the same as 20/40. If the examining room is not twenty feet long, the eye doctor utilizes a series of mirrors or a special projector to calibrate the letters to an equivalent size.

As will be discussed in the section on presbyopia in chapter 12, near vision does not always enjoy the same level of acuity as distant. To see something at close range, one must focus the lens within the eye. This focusing ability diminishes with age. At about middle age the normal eye requires reading glasses, and these are gradually strengthened as the eye's focusing power declines. But given the appropriate near glass for one's age, near vision will be equal to distant vision. A young child has enormous focusing capacity. A child can actually compensate for diminished vision and see disproportionately well at close range simply by holding material closer to gain magnification. The opposite is true of the normal, older presbyope who, without appropriate reading glasses, must hold things farther away to focus on them. Near vision will be examined in individuals who are middle aged or older and in those whose vision is significantly reduced.

Visual Field Testing

For purposes of a general evaluation most eye doctors test the extent of the visual fields by confrontation techniques or with a visual field screening device. In the confrontation technique an examiner is positioned in front of and facing the patient, who stares at him with one eye covered. A hand or an object is brought into view from the periphery, and the patient responds as soon as it is visible. There are also various devices that screen the extent of the field and pinpoint nonseeing areas closer into the central field. If such screening techniques suggest any abnormality, or if the clinical history or a suspected disease process warrants it, formal field plotting will be done. Visual field plotting is discussed in the chapter on glaucoma but is applicable to any process affecting the fields. Other conditions besides glaucoma that can cause field loss are cerebral vascular disease, multiple sclerosis, other neurological diseases affecting the visual pathways, head trauma, brain tumors, aneurysms, abscesses, and retinal diseases.

Refraction and Glasses

While the level of uncorrected vision may be of obvious concern, only the best *corrected* vision is considered significant in evaluating for eye disease or visual limitation. As will be explained in the chapter on refraction, only in rare instances is the refractive error so severe as to constitute a real disease or abnormality, and it is seldom secondary to an ocular or systemic disease process. The eye doctor will check the patient's present spectacles to determine what corrective lens will provide the best possible vision. This best corrected vision is considered one's true level of vision.

To help decide upon the best corrective lens, the eye doctor often asks the patient to compare various lenses and to report which is better or worse than another. This technique, called *subjective refraction,* relies on the patient to refine the correction. There are no right or wrong answers on this test. Many alternatives or pairs are presented, but not necessarily in the same order each time, so the patient should not second-guess or worry that answers are inconsistent. The eye doctor discerns patterns of consistency and interprets them as an indicator of either the true level of vision or the patient's understanding of the test. The patient is asked to comment only on the overall clarity of the figures and sharpness of the details. Watching for the roundness of o's and c's or for details in other letters might make the choices easier, but the only correct answer is the lens with which the patient sees comfortably and well. It does no good to ponder each alternative and its effects on the figures. Such efforts require focusing of the eye, which alters the effect of the trial lenses. A quick response without any effort or focusing is preferable. This part of the exam can be performed with single trial lenses that fit into a frame placed on the patient or with a larger instrument filled with lenses (see figures 2–2 and 2–3).

Focusing is difficult or impossible to control in some older farsighted individuals who can no longer relax the continuous focusing they have used for many years to see clearly and in children and adults whose attention and fixation on a distant object cannot be maintained. In these cases, eyedrops called *cycloplegics* are used to paralyze temporarily the ability of the ciliary muscle within the eye to do this focusing. Besides paralyzing the focusing ability, these drops dilate the pupils and make the individual very sensitive to light. The most commonly used drops last only several hours, but there is great individual variation. Younger patients require stronger eyedrops to obtain a full effect, which then takes longer to wear off. The effects of atropine, a drop sometimes used with very young children, may last for up to a week or more in rare instances. It is always wise to bring dark glasses and arrange for someone else to drive if one anticipates that the eye doctor will administer such drops. The same or similar short-acting drops are used to dilate the pupil for a complete retinal examination, as we will discuss later.

As was mentioned, a full eye examination includes an assessment of the general health of the patient, for innumerable systemic problems and medications can have an effect on the eye or vision. Thus, the eye doctor notes any possibly significant abnormalities in the patient's appearance, skin, and general status.

The movements of the eyes, their alignment, and the muscle balance of the two together are examined by having the patient fixate lights or targets close up and in the distance. Steady fixation is important, so many eye doctors have special targets designed to hold the attention of children. Various dissociative tests gauge the natural tendency of the eyes to deviate, an important consideration

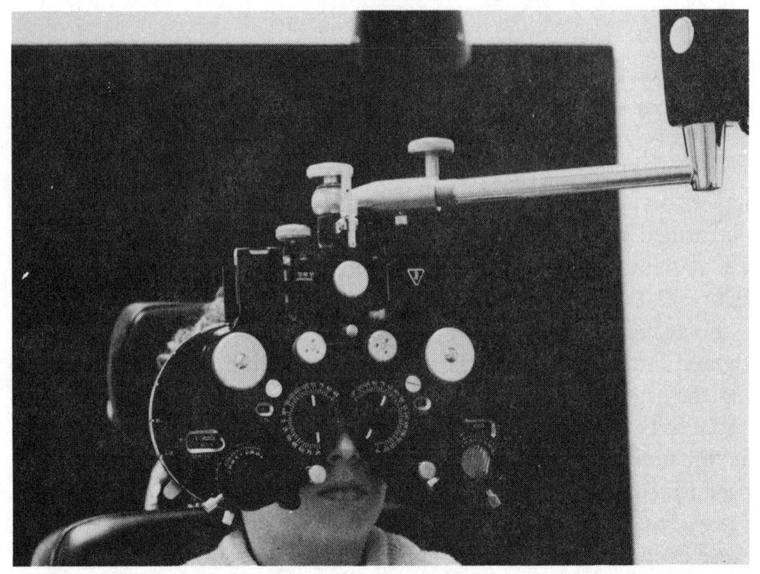

Fig. 2-2. Refraction, or determination of corrective lens power necessary for best possible vision, is measured by an instrument that contains an array of spherical and cylindrical lenses.

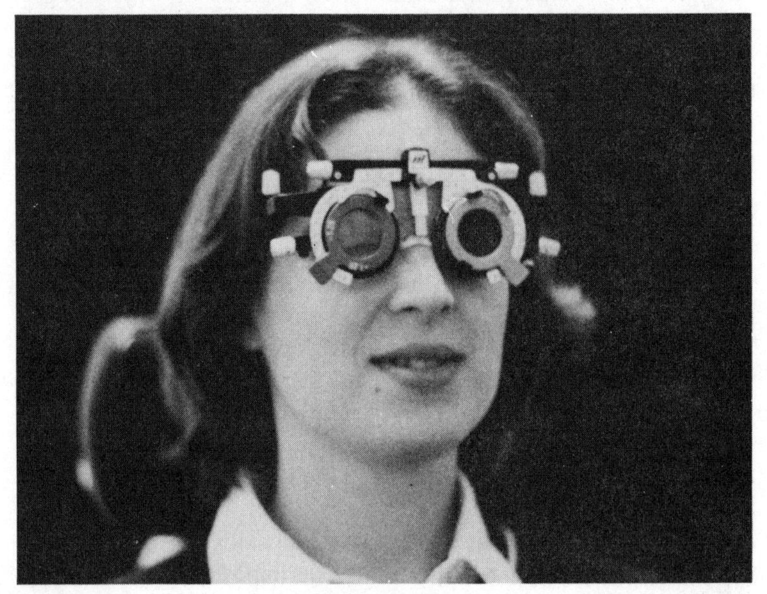

Fig. 2-3. A trial frame fitted on the patient may also be used to determine the refractive error. Separate trial lenses can be fitted in the frame, interchanged, and adjusted as necessary.

in differentiating types of eye strain. Although a tendency for a large ocular deviation, called a *phoria* (see chapter 7), can be compensated for with effort, this produces fatigue. These dissociative tests involve presenting a disparate image to each eye so that the brain will not fight double vision if an eye drifts out of line. For example, a colored glass lens or a special lens that transforms a circle of light into a line placed over one eye will dissociate the eyes and allow a measurement of the tendency toward deviation. A certain small deviation under these conditions is normal, but a significant, symptomatic deficiency of eye coordination may represent a problem that can be corrected with glasses or exercises.

Similarly, tests exist for evaluating the *fusion*, or use of the two eyes together. Manifest deviations of the eye, whether constant, intermittent, or latent, may impair the functional use of the eyes together. The ultimate measure of binocularity is *stereopsis*, three-dimensional vision. The simplest tests for three-dimensional vision involve the use of polarized glasses to discern smaller and smaller angular discrepancies, which will appear to the viewer as lesser amounts of three-dimensional displacement of the objects from the page. In this way, stereopsis can be evaluated and quantified. Chapter 7 on strabismus and amblyopia discusses the significance of impaired three-dimensional vision.

In evaluating the movements of the eyes, the reaction of the pupils to light and accommodation, the action of the lids, and corneal sensitivity, the doctor tests the function of six of the twelve cranial nerves, which are the major direct nerves from the brain. Problems affecting the brain and nervous system may also affect or present as abnormalities of eye movement or coordination, pupillary abnormalities, changes in the sensation in and about the eye, or diminished eyelid strength. Aneurysms, tumors, abscesses, meningitis, encephalitis, trauma, inflammation of nerves, and diseases such as multiple sclerosis all have ocular consequences.

While most of the preceding external examination is conducted by observation or with simple handlights, the remainder of the eye examination requires special instruments. Much of the eye doctor's equipment may look as forbidding as the dentist's, but despite all the instruments and gadgets, no part of even the most comprehensive eye examination is painful. The worst one can expect is some momentary discomfort from lights shining in the eyes.

Instruments

Slit Lamp

Figure 2–4 is a photograph of a typical slit lamp. This ingenious device consists of a binocular microscope adapted to a movable light source that can vary in shape from a slit beam to a larger band of illumination. The examiner can thus see a variably magnified eye in three dimensions. A specific advantage of the slit lamp is that it can use a tangential slit, or slice, of illumination to examine the eye's component parts at varying depths. The instrument can, for example, ascertain the depth of a corneal laceration or foreign body.

The slit lamp is used routinely for the magnified examination of lids, conjunctiva, cornea, anterior chamber, iris, lens, and anterior vitreous. It reveals in microscopic detail such abnormalities as early lid infections, tumors, conjunctivitis, minute foreign bodies, corneal infections or abnormalities, inflammations or infections within the eye, cataracts, and vitreous abnormalities. With special lenses it can also be used for a magnified, three-dimensional examination of the retina, optic nerve, and vitreous, as well as the anterior chamber filtration angle. This latter exam, called *gonioscopy,* is described in chapter 5.

Ophthalmoscope

The inside of the eye—the optic nerve, retina, and choroid—is examined with an ophthalmoscope. By enabling the doctor to look in through the pupil to the *fundus* of the eye, as the inside is called, the direct ophthalmoscope (the type most commonly used) provides a fascinating view of the optic nerve, the blood vessels, and retina (figure 2–5). The fundus is the most informative part of the body: only from here can nervous tissue and blood vessels be seen so directly. Examination of the optic nerve head, which is seen directly, can disclose a wide range of problems, from glaucoma to brain tumors. The blood vessels of the retina may show signs of high blood pressure, arteriosclerosis (hardening of the arteries), inflammation, or blood-borne infections. The fundus

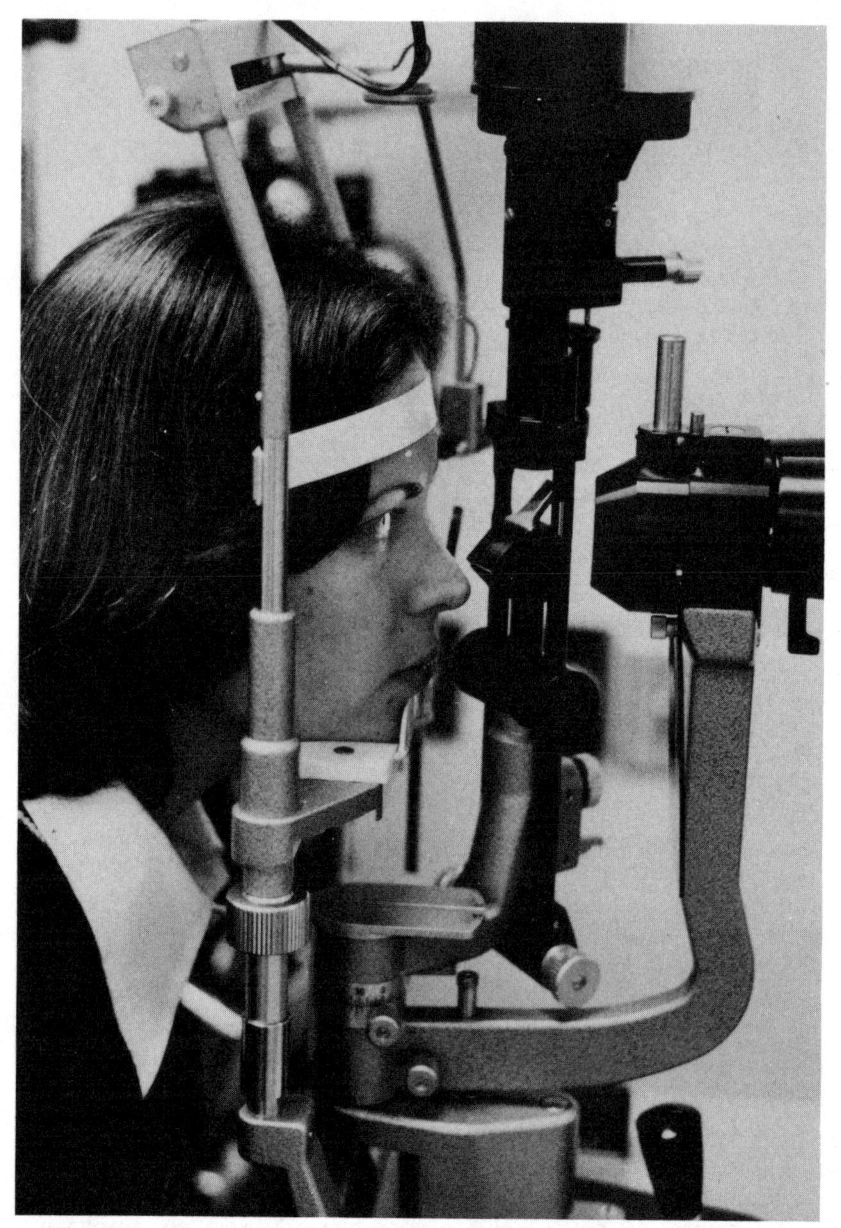

Fig. 2-4. The slit lamp affords the examiner a stereoscopic, magnified view of the surface of the eye and its internal structures.

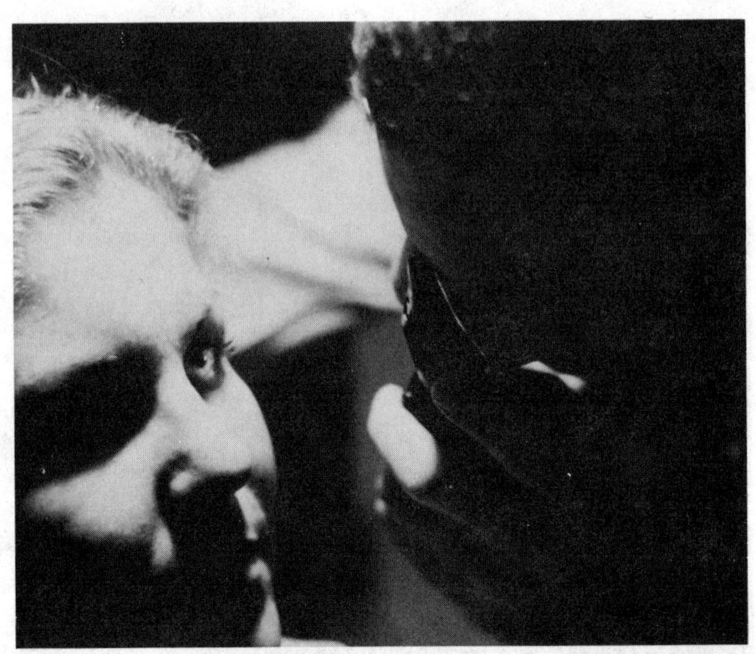

Fig. 2-5. The direct ophthalmoscope is a hand-held instrument used to examine the retina and optic nerve inside the eye.

of the eye can reveal problems that affect the body elsewhere as a whole, as well as those that are unique to the eye, such as glaucoma, retinal degenerations like night blindness, macular degeneration, retinal tears and detachments, and tumors. These ailments are discussed at length in their specific chapters, and the ocular effects of the more common systemic diseases are described in chapter 11.

A complete examination of the retina will include the periphery (the anterior part), which can be examined only by administering drops to dilate the pupil. The drops may take up to several hours to wear off, but rarely will their effect persist for more than a day. One might experience blurred vision, especially at close range, and extreme sensitivity to light; so anyone who anticipates having the pupils dilated as part of an eye exam should bring dark glasses and arrange for someone else to drive home. Most often, no blurring occurs for distance vision and on a dark or overcast day the light sensitivity is minimal. But it is best to be prepared.

Examining the eye through the dilated pupil is like looking into a room through an open door rather than through the keyhole—it affords a much better view. A retinal tear or early detachment is most likely to occur in the periphery, so dilation is needed for detection. The instrument used most commonly for three-dimensional examination of the retina out to its anterior periphery is the binocular indirect ophthalmoscope. With this ingenious contraption on, the ophthalmologist looks like a miner (figure 2–6). To

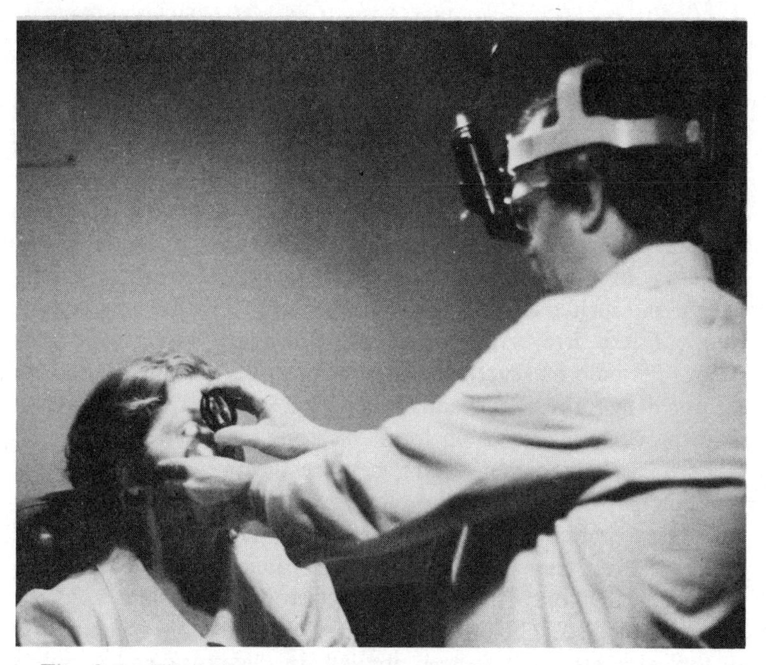

Fig. 2-6. The indirect ophthalmoscope affords a stereoscopic, panoramic view of the retina and the inside of the eye.

examine the extreme periphery, he may use *scleral depression* in conjunction with the scope. This technique consists of putting gentle pressure around the lids and outside of the eye in order to bring the most anterior edge of the retina into view. While the intensity of the light from this scope may make the patient uncomfortable, it is painless and harmless when used with or without scleral depression. The patient should relax and look in the direction requested, keeping both eyes open at all times (except for necessary momentary blinking). The eye not being examined will serve to fixate objects in the desired direction.

With this examination ophthalmologists can diagnose retinal tears or very early detachments at a stage amenable to much simpler therapy than the major operation needed to repair a large retinal separation. Small tumors and other abnormalities not noticeable through an undilated pupil also may be discovered.

The Glaucoma Test

Glaucoma testing is discussed more fully in chapter 5. We will mention here only that this test is routinely administered to all people over forty and to anyone whose history, examination, or previous tests arouse the doctor's suspicions. The test measures the pressure within the eyeball, as glaucoma is a disease of abnormally high pressure in the eye. The most common variety slowly and insidiously destroys the optic nerve until 90 percent or more of it is irreversibly damaged before any visual symptoms occur. Periodic measurements of the pressure within the eye are an excellent form of preventive medicine. Therapy will almost always prevent loss of vision from this common disease.

Fortunately, tonometers can measure the pressure very simply and quickly with no danger to the eye. The air tonometer measures the pressure after a calibrated momentary gush of air is reflected from the eye. It is a good screening instrument. Figure 2–7 illustrates the Schiotz indentation tonometer, commonly used by ophthalmologists. The instrument is set on the eye, to which a drop of topical anesthetic has been applied. In this quick and harmless test, the minimal amount of indentation caused by the calibrated weight is read on the scale and translated into a pressure reading. Another accurate tonometer is the applanation device. It is most common in ophthalmologists' offices, for the usual type requires a slit lamp. With this device, topical anesthetic and a yellow fluorescein dye (a simple eyedrop) are placed in the eye. A tiny

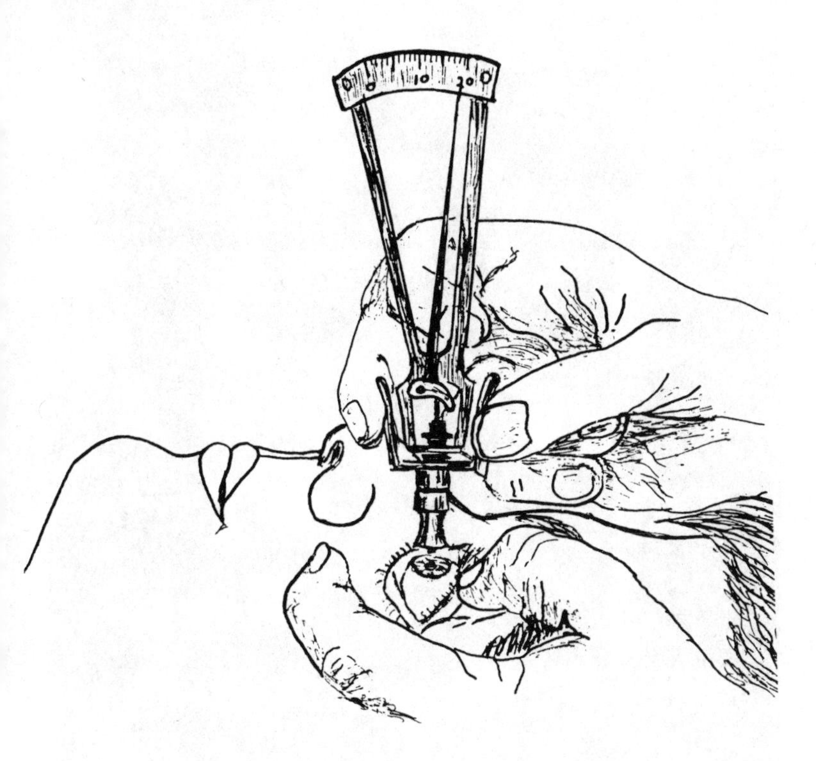

Fig. 2-7. The Schiotz tonometer is used to measure the pressure in the eye. It is a quick, simple, and painless test for glaucoma.

amount of corneal flattening results in a fluorescent ring meniscus seen by the ophthalmologist, and the pressure exerted by a spring gauge to cause this flattening is read directly as the intraocular pressure (figure 2–8).

The eyes are temporarily anesthetized by the drops so that the patient does not feel the momentary touch of either instrument. Since the lids are not anesthetized, a blink will of course make one aware that the instrument is touching the lid then. But no harm comes from the test; the harm is in avoiding it. Because the eyes are anesthetized, one should not rub them. Tears or excess drops can be blotted away, but rubbing might scratch the eye while the

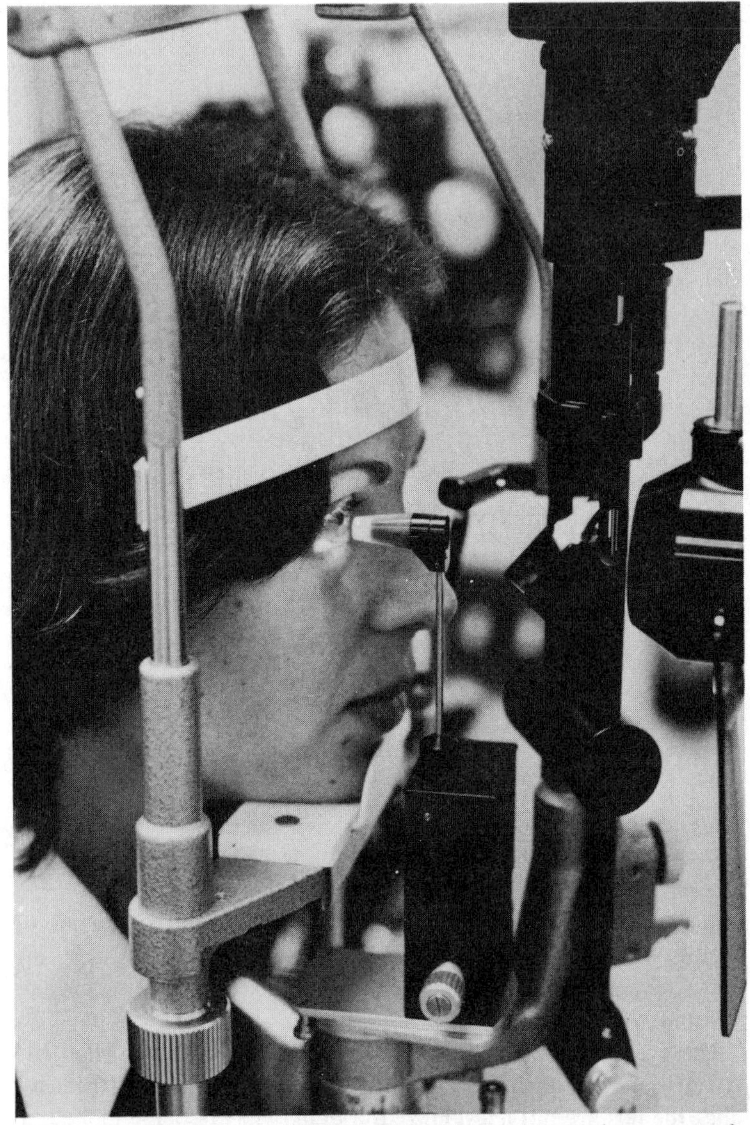

Fig. 2-8. The applanation tonometer is another device commonly used to measure the pressure within the eye. This measurement is also brief and painless, since topical anesthetic drops are applied to the eye before the measurement is taken.

sensation is not present. The effect of these drops—a numb, heavy feeling—dissipates within a few minutes.

The tests and procedures reviewed in this chapter are used in most routine eye examinations. Many specialized examinations are performed, and doctors use different techniques and instruments, but what counts is that the exam be thorough enough to rule out ocular disease, systemic abnormality, and ocular side effects of systemic medications or diseases. If difficulties or questions arise after a prescription for glasses is filled, the doctor should be asked to check the accuracy of the prescription and the glasses.

3
Brief Anatomy and Physiology of the Eye and Visual System

Most discussions of the anatomy and functioning of the eye compare it to a camera. Both eye and camera focus images on a recording device, and both do so in a very similar fashion with roughly analogous parts. In fact, observations on the functioning of the eye often antedated similar discoveries in photography. Of course, in the case of a camera the image is recorded on film as a momentary and stationary scene, whereas in the eye information is picked up by the retina, the "film" of the eye, and transmitted immediately via the optic nerve to the brain, where we perceive ongoing visual stimulation.

We see because all objects reflect a portion of the light that hits them. Reflected light rays enter the eye and stimulate the visual system. The cornea—the round window, or crystal, of the eye—does most of the focusing. It is approximately 12 millimeters (½ inch) wide and normally bulges forward like a portion of a sphere. Its regular and definite spherical shape enables the cornea to gather and focus light rays, but it cannot adjust the focus.

Without a change of focus an object seen clearly in the distance would suddenly become blurred as it approached because the light rays reflected from nearby strike the eye at a much different angle than do the ones from distant objects (see figure 3–1). Therefore,

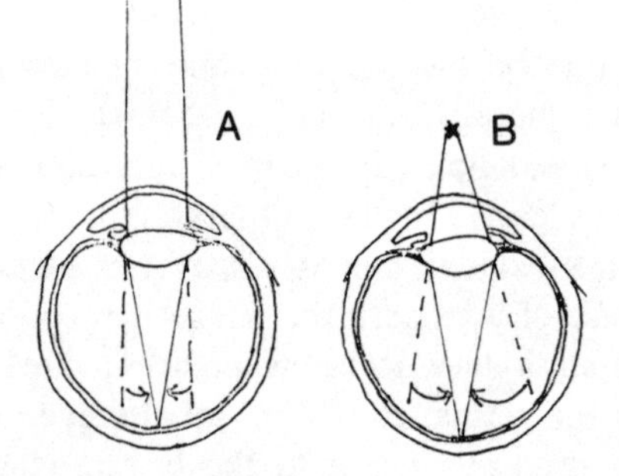

Fig. 3-1. An object seen clearly in the distance is focused upon the retina (A). When an object is closer to the eye, light rays from the object enter the eye at a more widely divergent angle. These must be gathered with increased focusing power by the eye in order to project a clear image on the retina (B).

in order to see well both at close range and in the distance, the eye must be able to modify its focusing power. This variability is provided by a lens in the eye which is simply (unusual for medical terminology) called the *lens*. This lens is a clear, crystalline, convex structure that sits directly behind the pupil, the black spot that you can see in the center of the eye. The pupil is a round hole through which the light rays enter the eyeball.

The lens within the eye varies its focus by an interesting process.

The lens is suspended directly behind the pupil by very fine thread-like strands called *zonules*. These in turn connect to a circumferential ciliary muscle. When the normal eye focuses in the distance, the ciliary muscle is relaxed. In this position—its least convex and thus least powerful configuration—it stretches the zonules and keeps the lens taut. Focusing close up requires greater optical, or focusing power, so the ciliary muscle contracts and rides forward, loosening the zonular ligaments which then allow the natural elasticity of the lens to form a thicker, more convex shape. This configuration enables the lens to gather the more divergent light rays that reflect from a nearer object (figure 3–2).

As we age, the elasticity of the lens decreases, thus weakening the focusing power. This gradual loss of elasticity and focusing power is a reliable index of age. Focusing ability is very great in early childhood and then diminishes substantially until it becomes almost negligible at about age seventy. At about age forty-five, otherwise normally sighted people may require reading glasses to compensate for the loss of focusing power. This phenomenon will be explored in chapter 12.

The colored portion of the eye surrounding the pupil is the *iris*. This structure fulfills the same function as the iris diaphragm in a camera. The iris dilates to make the pupil larger, thus allowing more light to enter the eye in poorly illuminated areas; the iris also constricts to make the pupil smaller, thus shielding the eye from excessive light and glare in very bright areas. Anyone who has had drops that dilate the pupil for a time can appreciate the remarkable ability of the pupil to constrict automatically. Out in the bright sunlight, one is uncomfortable because the drops have prevented the pupil from constricting to a more comfortable small size appropriate to the bright surroundings. More commonly, one notices the effect of this pupillary constriction upon going from a very dark room out into the bright sunlight. The pupil remains momentarily dilated, producing a sudden and uncomfortable dazzling effect. This is a normal sensation and is no cause for alarm, but persistent sensitivity to light, called *photophobia*, can indicate certain disease processes or abnormalities which do deserve medical attention.

Between the iris, the colored part of the eye, and the cornea, the

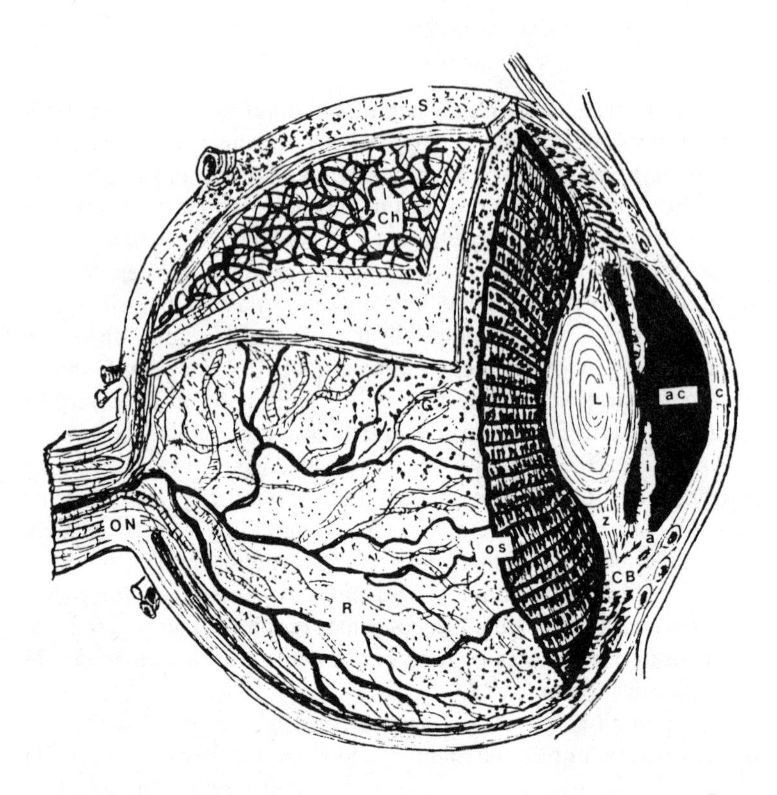

Fig. 3-2. The anatomy of the eye from the side. The cornea (c) is the front crystal of the eye overlying the anterior chamber (ac). The colored iris (i) is the diaphragm regulating the size of the black pupil through which light rays enter the posterior chamber of the eye. The lens (L) provides variable focusing power for these light rays and is suspended from the ciliary body (CB) by its zonular ligaments (z). Aqueous humor is produced by the ciliary body and nourishes the lens and inner cornea, flowing through the pupil into the anterior chamber and out through the filtration angle (a) formed at the junction of the cornea and base of the iris. The retina (R) lines the posterior two-thirds of the eye, attached anteriorly at the ora serrata (os) and posteriorly at the optic nerve (ON). Between the retina and the tough white outer coat of the eye called the sclera (S) is the intermediate vascular, pigmented layer called the choroid (Ch). It nourishes the outer retinal layers.

crystal of the eye, is a clear fluid called the *aqueous humor*. This fluid circulates in the eye to transport nutrients to the lens and inner cornea. It also carries away waste products of metabolism. The eye needs this clear fluid circulation because blood and blood vessels would render structures opaque. The fluid is produced in the ciliary body directly behind the iris; it then flows through the pupil into the anterior chamber and out through the trabecular meshwork located in the junction, or angle, formed between the iris and the cornea (figure 3–3). After exiting through this mesh-

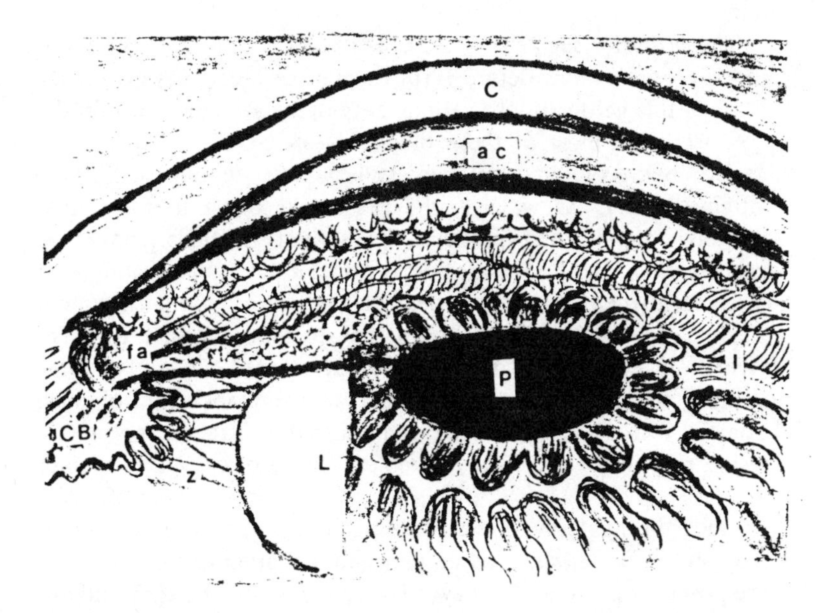

Fig. 3-3. A detailed drawing of the anterior segment of the eye showing the cornea (C), anterior chamber (ac), iris (I), pupil (P), filtration angle (fa), ciliary body (CB) and muscle, and a portion of the lens (L) and its suspensory zonules (z) through the cutaway segment of iris.

work, the fluid enters a larger channel called the *Canal of Schlemm*. It then drains into small veins in the outer coats of the eye. The lens of the eye is a living, active tissue, then, needing nutrition and producing wastes. Disorders of the lens or its metabolism can interfere with its ability to focus or cause it to become cloudy, a condition known as cataract (see chapter 6).

Thus, we can control the light rays entering the eye just as we control them in a camera. The light rays are gathered and focused by the cornea at the front of the eyeball. The correct amount of light is then allowed to enter the eye by having the pupil vary with the dilating or constricting muscles of the iris. Fine focus is achieved by adjusting the shape of the lens. The light rays entering the dark chamber of the eye posterior to the pupil then proceed back to a focus on the retina, just as light rays go through the posterior black chamber of the camera to be focused upon the film.

The posterior segment of the eye is filled with a clear gel called the *vitreous humor.* Before birth, numerous blood vessels course through this gel to supply various developing structures within the eye. At birth, most of these blood vessels have already disintegrated, but tiny remnants may still be left floating within the vitreous cavity. As we age, the gel breaks down to form fluid pockets, strands, and particles. These imperfections appear as *vitreous floaters*—lines and spots drifting about in our vision, most often seen when staring at a clear blue sky or a plain background. Seeing a few of these spots from time to time almost always indicates no eye disease or problem. The sudden onset of a large number of such floaters, however, or the association of such floaters with flashes of light and blurred central or peripheral vision may indicate a serious problem, to be discussed in chapter 8.

Once the light rays have coursed through the vitreous humor, they strike the *retina,* a tissue paper-thin coating on the inside of the posterior portion of the eye. The retina is composed of millions of tiny light-sensitive cells. When light hits them, these cells generate a chemical reaction which in turn stimulates an electrical reaction in the nerve fibers of the retina. Electrical impulses traveling from the nerve fibers in the retina to the optic nerve and then back to the brain constitute what we finally perceive as vision.

The structure of the retina is truly remarkable. It contains approximately 130 million rods and 7 million cones, tiny cells that perceive light. The rods react to visible light only, irrespective of its wavelength. These cells thus allow us to see the difference between light and dark, but they do not distinguish colors. Visible light is a spectrum of wavelengths, with each perceived as a definite color. Colors are distinguished by the cones, cells that look different from the rods and react specifically to a given wavelength in the visible light spectrum. At present there is good evidence to suggest that there are three different types of cones representing red, green, and blue. Stimulating various combinations of these cones allows us to see the whole spectrum of colors we know. Abnormalities of the cones, arising from inheritance or disease, cause various abnormalities of our color vision. Color blindness and related problems are discussed in chapter 8.

The rods are far more sensitive to low levels of light stimulation, but the cones allow us to see more discretely and exactly. At night, when things are very poorly illuminated, we are using the rods in our retina, so we do not appreciate colors as well, nor are we able to see fine detail. We see most keenly with the cones because some are set up in a one-to-one relationship with individual nerve fibers of the optic nerve, concentrated in a very regular array at the posterior pole of the eyeball, where the directly focused light rays will fall. Rods, on the other hand, are located primarily in the periphery and have only one nerve fiber to serve several. Thus, you can understand why an object can be seen better in the dark by looking slightly to the side of it. In so doing, one is focusing on the rods of the retina rather than the cones.

The tiny central portion of the posterior pole of the eye, which gives us the best detailed vision under lighted conditions, is composed almost entirely of cones in a one-to-one relationship to the nerve fibers. This area is called the *fovea.* It is only about 0.5 millimeter in diameter. Surrounding the fovea is the *macula,* an area of slightly less discrete connections to the optic nerve, but with reasonably good acuity. The whole macular area, which includes the central fovea, is approximately 3 millimeters in diameter. It enables one to see very fine print such as the 20–20 line on the eye doctor's chart. Any disease or process that destroys the function of this tiny pinhead of retina at the back of the eye usually reduces vision to the extent that only the large E at the top of the chart is visible.

The nerve fibers emanating from the rods and cones travel in a very orderly fashion along the innermost layer of the retina into the optic nerve. The nerve is located a few millimeters away from the macula. It is on the nasal (closer to the nose) side of each eye. From the optic nerve head these fibers pass out the back of the eyeball, like a cable from a TV camera, transmitting what is seen to the brain. The optic nerve head, which is visible when the ophthalmologist looks inside the eye, is only about 2 millimeters in diameter. Because there are no receptor cells (rods and cones) overlying it, the position of the optic nerve head constitutes a blind spot in the visual field. This phenomenon is illustrated in figure 3–4. Visual field is the term for the entire area encompassed

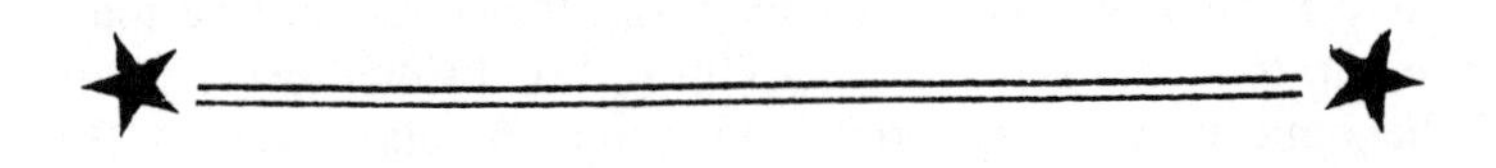

Fig. 3-4. To demonstrate the normal blind spot, close the left eye and focus the right eye on the star on the left. When the page is moved approximately 10 inches from the head, while looking directly at the left hand star the right hand star will disappear. When the left eye is focused on the right hand star at ten inches from the page, the star on the left will disappear.

by the vision of the eye. The eye chart with letters measures only visual acuity, the fine central vision governed by the macula and fovea. The visual field is the total area in which one perceives light and objects. Figure 3–5 shows the optic nerve head, the blood vessels that come from the optic nerve and run along the surface of the retina, the retina itself, and the macular and foveal area as visible when the eye doctor looks through the pupil with an ophthalmoscope.

The two optic nerves exit the eye sockets through small bony canals; then, shortly after entering the bony cavity of the skull that encases the brain, they partially cross over. Half of the fibers of

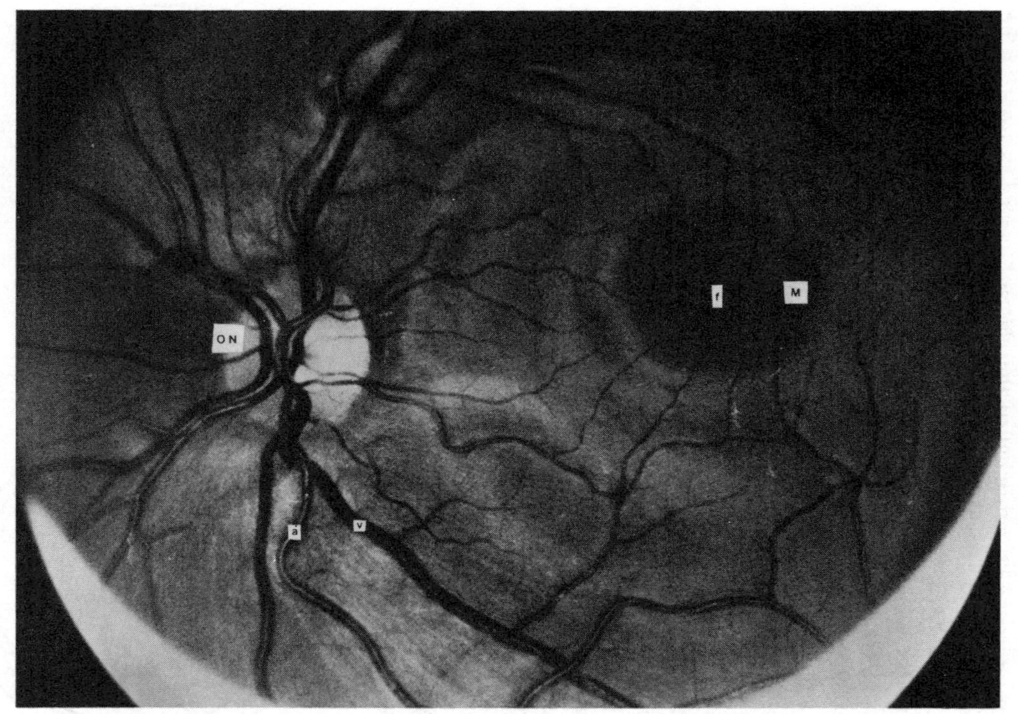

Fig. 3-5. The posterior pole of the eye as seen by the doctor through an ophthalmoscope. One can see the optic nerve head (ON), the retinal arteries (a.) and veins (v), the macular area (M), and the fovea (f).

each nerve, representing the nasal half of each retina, cross over to the opposite side of the brain. This crossover of the two optic nerves forms a structure called the *chiasm,* located in front of the pituitary gland. Posterior to the chiasm, the fibers of the visual pathways transmit the visual field of the opposite side. To understand the exact relation between the visual fields and the disposition of these nerve fibers, one must understand first that the eye, like a camera, has an optical system which projects all images upside down and backwards on the retina (figure 3–6a). Thus, behind the chiasm on the right side of the brain one finds the fibers from the nasal half of the left eye's retina and the temporal (outer) half of the right eye's retina. These fibers transmit the left side of the vision from each eye. Similarly, on the left side of the brain are the fibers serving the right half of the visual field in both eyes. Along the superior (upper) portion on both sides of the brain are the fibers serving the lower halves of our vision in each eye, and along the inferior portion are the fibers serving the upper half. A simple representation of these relationships is seen in figure 3–6b. You can now understand why examination of the visual fields is an important step in discovering the effects of various diseases not only on the eye, the retina, and the optic nerve, but also on structures such as the pituitary gland and the brain. These

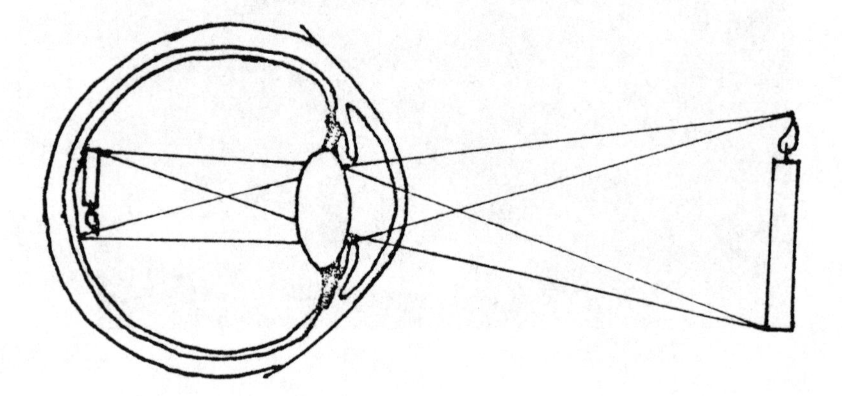

Fig. 3-6A. Each point viewed in space reflects or emits light rays, which are focused by the optical sysem of the eye upon the retina so as to project the image upside down and backward.

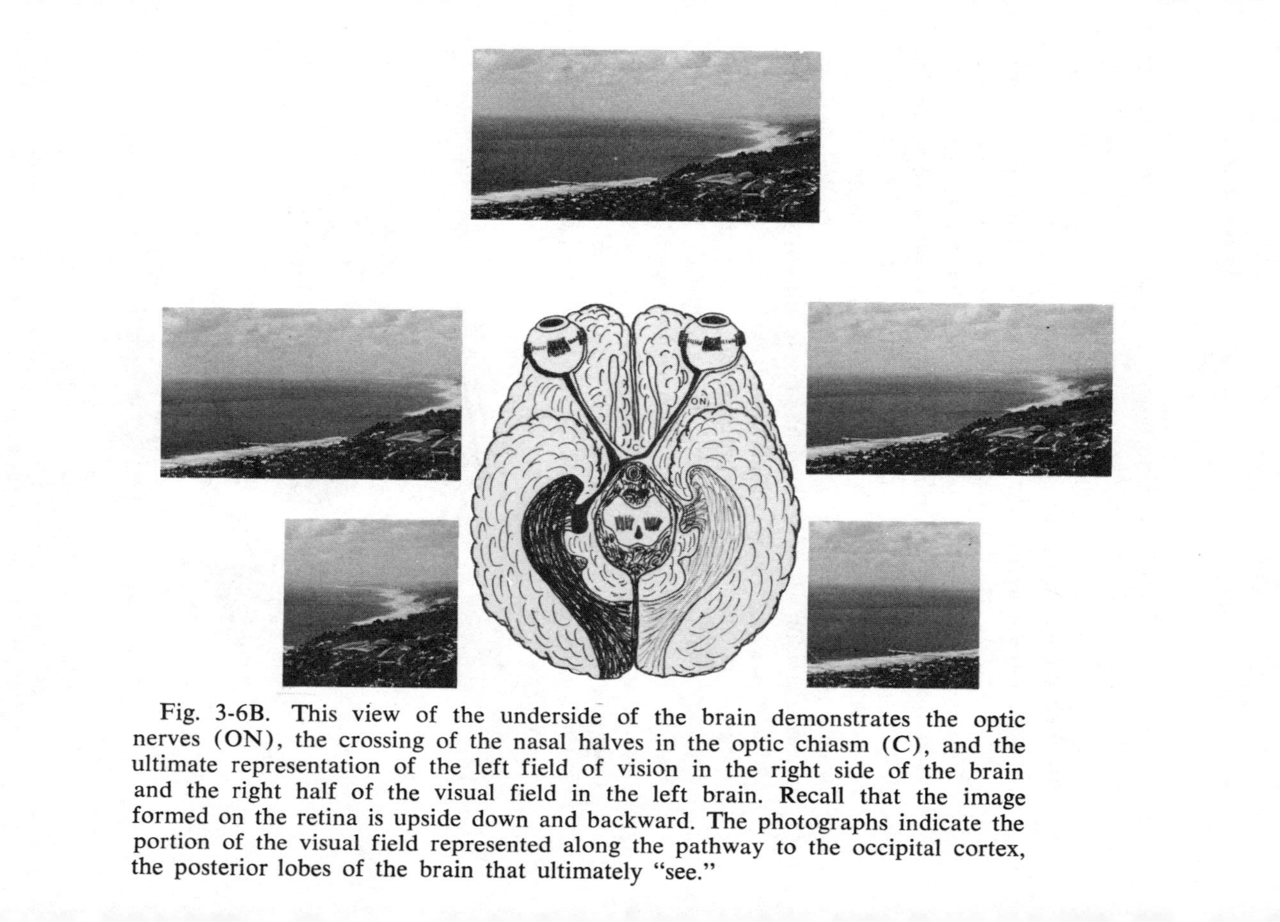

Fig. 3-6B. This view of the underside of the brain demonstrates the optic nerves (ON), the crossing of the nasal halves in the optic chiasm (C), and the ultimate representation of the left field of vision in the right side of the brain and the right half of the visual field in the left brain. Recall that the image formed on the retina is upside down and backward. The photographs indicate the portion of the visual field represented along the pathway to the occipital cortex, the posterior lobes of the brain that ultimately "see."

relationships and common disorders that disrupt them will be studied more closely in chapter 11.

The eyeball itself, with the exception of the 12 millimeters in the front covered by the clear cornea, is encompassed by a thick white fibrous coat called the *sclera.* Between the sclera and the retina is a dark, highly vascular coat of the eye called the *choroid.* The dense pigmentation in the choroid helps to eliminate light from the posterior portion of the eye, while the rich blood supply provides nutrition for the outer layers of the retina. Because the choroid is so vascular, it can succumb to blood-borne infections and diseases. Unless a biopsy (surgical excision of a piece of the body's tissue for examination) is performed, the eye is often the only place that the doctor can directly visualize such widespread processes.

Surely the most common problems presenting themselves in the eye doctor's office are those involving the eyelids and conjunctiva. Eye doctors call these disorders external diseases. The eyelids are delicate structures, covered by an outer layer of skin and coated on the inside by conjunctiva, a thin mucous membrane that above and below forms a cul-de-sac, or dead end, and folds back onto

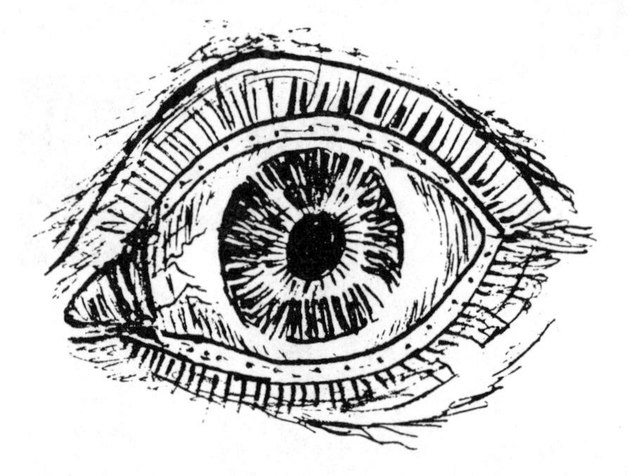

Fig. 3-7. The puncta are the white, often slightly elevated, tiny openings at the inner end of each lid through which the tears exit. The orifices of the Meibomian glands form a line directly behind the lashes of each lid.

the eyeball itself. The only external portion of the eye not covered by this conjunctiva is the cornea, or crystal of the eye, through which the colored iris and pupil are visible.

The inner substance of the eyelid is composed of a thick fibrous tissue, called the *tarsus,* which maintains its shape. In front of the tarsus is a muscular layer which allows one to squeeze, close, and blink the eye. At the outer edge of the eyelid margin there is normally a single row of eyelashes, all of which should point outward to avoid scratching and irritating the eye. Immediately behind the row of lashes one can see with magnification a single row of tiny openings from the small oil glands within the eyelid (figure 3–7). These glands, called the *Meibomian glands,* produce an oily substance which is the outermost component of the tear film that covers the eye. It prevents rapid evaporation of the watery inner tears yet keeps them from running over the lid margin. Infections in the Meibomian glands are commonly called styes. Chapter 4 says more about disorders of the lids and Meibomian glands.

The *conjunctiva* is a thin mucous membrane containing cells

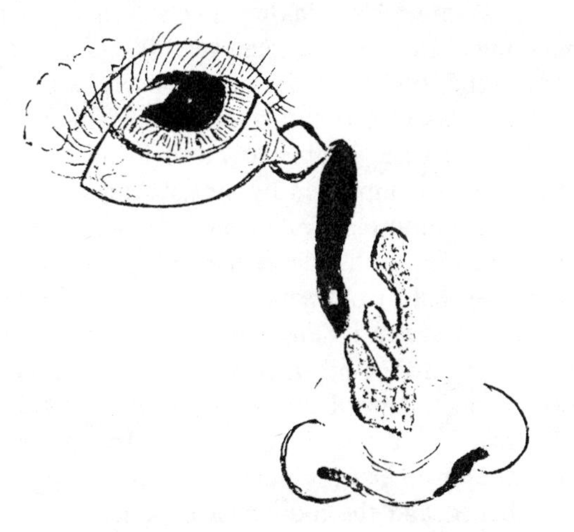

Fig. 3-8. The tear gland is located at the upper outer corner of the orbit. Tears flow across the surface of the eye into the puncta at the inner corner, through the canaliculi and into the tear sac, then down the nasolacrimal duct into the back of the nose.

that contribute mucous to the tear film. Most of the watery component of the tear film is produced by small glands in the cul-de-sacs and by a large tear gland in the upper outer corner of the eye socket. When one cries or when the eye is irritated, watery tears flow from the large gland down across the eye. The tears exit through a hole at the inner corner of each lid. These holes, called the *puncta,* lead into tiny canals that join at the innermost corner of the lids against the nose, forming there the tear sac. This sac then drains into a small tube that goes through the bone and into the posterior portion of the nose (figure 3–8).

All the components of the tears must be present to maintain normal moisture in the eye. Without adequate tears the eye dries readily, the cornea loses its lustre, and vision suffers. It is also important that tears be able to drain out of the eye; otherwise they spill over the eyelids and create chronic tearing and wet, irritated lower lids. Obstructions in the small canals, called *canaliculi,* or in the lacrimal (tear) sac, or even down in the nose at the lower end of the so-called nasolacrimal duct, can all cause tearing. In addition, the lids' muscular blinking function must be present to produce pumping action on the tear sac. Without this muscular pump, the tears tend to well up and overflow just as from an actual obstruction. Disorders of this system considered in chapter 4.

The eyeball, of course, is able to move in all directions. This range of motion is accomplished by six muscles attached to each eye. The four *rectus* muscles, called the medial, lateral, superior, and inferior, attach on each side at the 12-, 3-, 6-, and 9-o'clock positions of the eyeball. Contraction of any one of these muscles causes the eye to move in its direction. In addition, there are two oblique muscles, superior and inferior, that wrap around the eye and allow it to rotate inward and outward around a longitudinal axis and to move into the oblique corners. The muscles in both eyes must operate together with great precision. If they do not, the eyes are not aligned and the individual experiences double vision instead of normal, binocular, three-dimensional vision. Children who are born with malaligned eyes, a condition called *cross-eye, wall eye,* or *strabismus,* usually develop adaptations to the malalignment of their eyes in order to avoid double vision. These disorders will be discussed in chapter 7.

4
External Diseases of the Eye

Mild inflammations of the outer lining of the eye, the conjunctiva, are extremely common. What many people fail to realize, however, is that a pink or red eye is only a sign that something is wrong; it is not a diagnosis in itself. There are many different conditions, some of them progressive and sight-threatening, which manifest themselves as redness, so it is dangerous to use nonprescription eyedrops that remove redness unless one knows why the redness is there and whether or not the drops are appropriate. All eye doctors have seen at least an occasional ill-advised patient who successfully relieved the redness with over-the-counter eyedrops, only to find later that an untreated disease had progressed to irreversible loss of vision. To be sure, most instances of mild and transient redness of the eye with minimal burning or watering represent simple allergies or irritations. They can appropriately be left untreated or treated with simple, nonprescription decongestants. But consider the host of other conditions that may present with only redness: acute glaucoma; significant lid, conjunctival, and corneal infections; severe allergic reactions; foreign bodies on the cornea or the conjunctiva; and inflammations within the eye itself that can lead to scarring of the pupil, glaucoma, cataracts, and retinal destruction. Usually such problems have additional symptoms sooner or later, but the spectrum of individual responses to these conditions is large enough to include many who experience only redness until further disastrous complications ensue. We will now consider in more detail the common conditions that affect the external portion of the eye: the lids, conjunctiva, cornea, and the lacrimal, or tear, system.

The Eyelids

As explained in chapter 3, the lids are composed of an outer layer of skin that is thinner and finer than in most other areas of the body, a layer of muscles that control blinking and closure, and a fibrous layer that gives the lids their firmness and contour. The skin of the lids is subject to almost all of the diseases and problems of skin elsewhere. It becomes easily inflamed and irritated from discharge, chronic excess tearing, and frequent rubbing. It is a common site for skin tumors, so any enlarging lump or sore that does not heal quickly should be examined by a physician (figure 4–1).

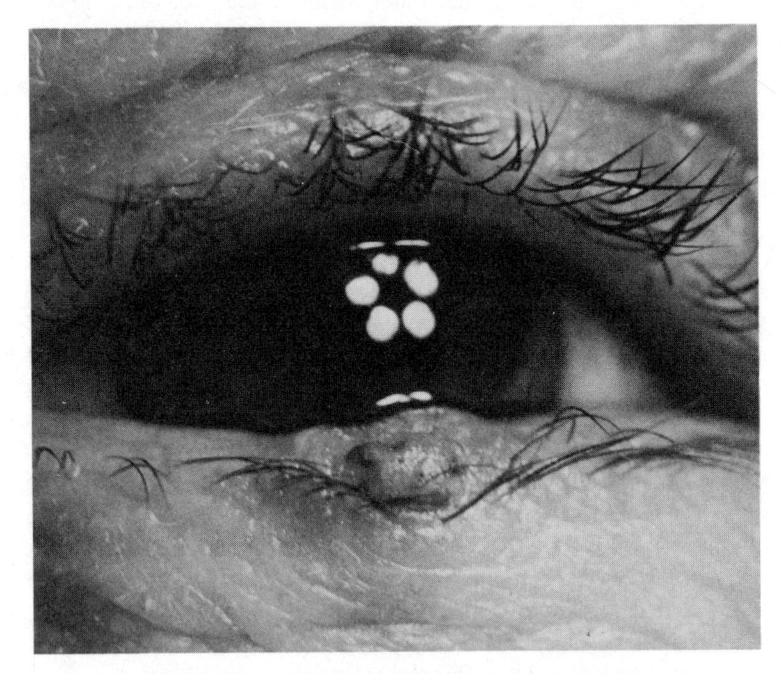

Fig. 4-1. A suspicious lesion of the eyelid margin.

Meibomianitis (*Stye, Hordeolum*)

A stye is a bacterial infection of the Meibomian glands in the tarsus, or firm middle layer of the lid (see chater 3). It is one of the most frequent eye complaints, commonly caused by the staphylococcus, or "staph," bacteria. The oil glands lined up along the lid are shaped like little goblets, with a thin neck opening up onto the lid margin. Infection easily plugs up this narrow outlet and the gland balloons up, forming a localized swelling in the lid. If the outlet is firmly sealed, the infection may point on the skin or conjunctival (inner) surface, forming a whitehead. Hot compresses are an effective means of opening the normal outlet, or the pointing area, to drain the active infection and debris. Antibiotic ointment or drops eradicate the bacteria that caused the infection in the first place.

Chalazion

The outlet to a gland sometimes becomes permanently scarred, either from a stye or from an unnoticed previous infection. The gland continues to produce oily material, and it starts to expand, producing a large lump in the lid. These lid cysts, called *chalazions,* are very common. If compresses and the passage of time do not bring improvement, the eye doctor can remove a chalazion with a simple office procedure. This minor operation usually leaves no scar, as the cyst can be removed from the inside surface of the lid. These cysts are not tumors and are not precancerous, but only a doctor can distinguish a typical chalazion from the rare suspicious lump deserving biopsy and examination for possible tumor.

Chronic Blepharitis (*Granulated Eyelids, Inflamed Eyelids*)

Chronic lid inflammations also are very common. The condition is usually caused either by a persistent staph infection of the oil glands of the lids or by seborrhea, excessive oiliness of the skin. In seborrhea, lid redness and flaking of the lashes are usually associated with a similar condition of the scalp, where it may appear as dandruff. This recurrent condition, called "granulated eyelids" in the past, is difficult and frustrating to treat. The doctor will prescribe medications as necessary for flare-ups and will give instructions on lid hygiene to try to minimize these episodes (figures 4–2, 4–3). Routine examinations are a must because complica-

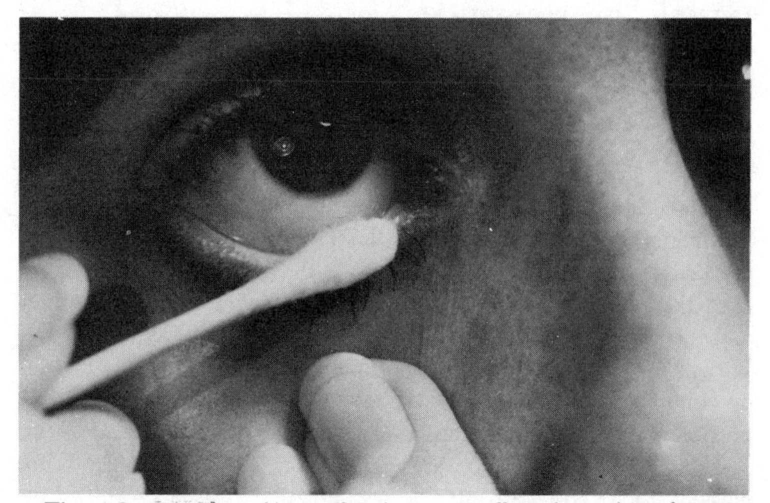

Fig. 4-2. Washing the eyelids is an excellent bacteriostatic measure for controlling chronic blepharitis. A cotton-tipped applicator soaked in a sudsy mixture of warm water and baby shampoo is used to wash along the bases of the lashes on the eyelid margin. One pulls down as far as possible on the lower lid with pressure against the bone below the orbit. Looking up as high as possible to rotate the cornea away, one can safely rub along the margin of the lower lid.

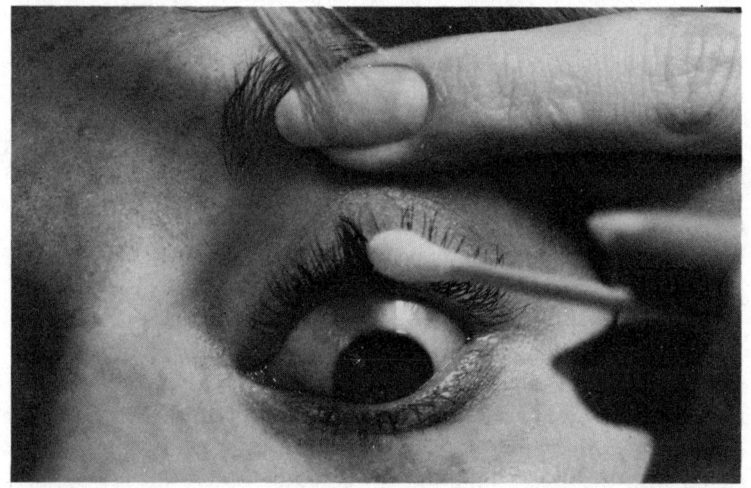

Fig. 4-3. To wash the upper eyelid, one pulls the upper lid upward and outward against the upper bony rim of the orbit. The eye looks down as far as possible to rotate the cornea away. The upper lid margin along the bases of the lashes can then be rubbed with the applicator.

tions include not only acute styes, but also more serious irritations and infections that can damage the cornea.

Ptosis (Drooping Eyelid)

Certain eyelid problems deserve special mention. Drooping of the upper lid, called *ptosis,* is a condition usually encountered in children as a congenital problem. The child is born with a deficiency of one or both upper eyelid muscles or the nerves to them. When such a problem develops later in life it is called *acquired ptosis*; when brought on by simple irritation of the eye it is referred to as *protective ptosis.* The condition is indecorously labeled *senile ptosis* when it arises from relaxation and weakening of the tissues with age. Ptosis may be part of a more complicated problem involving a major nerve from the brain that also controls focusing, constricts the pupil, and operates several muscles that move the eye around. The lid may also signal a problem affecting an altogether different nerve supply, the sympathetic nervous system. This system partially controls lid elevation, dilation of the pupil, the tone of arterial blood vessels, and sweating on each side of the face. Ptosis can thus suggest aneurysms, tumors, and strokes. Or the upper eyelid may droop as an early sign of a generalized muscle disease such as *myasthenia gravis*. Thus, "simple" drooping of the lid should be brought to the attention of the ophthalmologist, who will then investigate the nature of the ptosis.

Primary uncomplicated ptosis can be treated by any of a number of procedures designed to keep the lid elevated while the eyes are open and looking straight ahead. When the levator (the main muscle, which elevates the upper lid) is able to function at all, a procedure is used to shorten it, thereby strengthening it and pulling the lid up. If no function is present, bands of strong tissue, called fascia or collagen, may be used to suspend the lid from the brow. This sling operation then allows the lid to move up and down with motion of the brow. To a greater or lesser extent, all operations that raise the lid also result in a deficiency of closure. Forced closure, squeezing or consciously blinking, is usually complete, but the lids may actually remain very slightly open while sleeping. The cornea may thus lie exposed and dry out. In most people, the eyes tend to roll up in the head when the lids close, a reflex called *Bell's phenomenon.* When present, it indicates that such minimal deficiency of the lid closure will be of no consequence. When this reflex is absent, or when the eye movements are restricted by other muscle or nerve involvement, ptosis operations must be approached with great caution.

Entropion, Ectropion (Lax Lids, Turned Lids)

Aging is characterized by a relaxation and weakening of the skin and supportive tissues of the body. Fatty tissue in many areas also tends to diminish. With orbital and lid tissues, two distinctly different conditions can arise from this same process. The lids no longer rest securely against the eyeball, so they may either tilt in and rub the eye or tilt out and fall away from it. When the *orbicularis,* or circular muscle, slides up, it tilts the lid margin inward and results in a disorder called *entropion.* The condition is painful and dangerous, as the eyelashes constantly rub and abrade the cornea, causing tearing and rendering it more susceptible to infection. If the orbicularis muscle slides down, the lid will tilt out. The result is *ectropion,* a condition that also breeds infection. The lacrimal drainage will not function well because the punctum falls away from the eye and the muscles do not pump the sac well. A pool of stagnant tears forms and chronic tearing is the result of its overflow.

Fortunately, quick, simple, and very effective surgical procedures are available for these conditions. Under local anesthesia the lids can be tightened and repositioned for immediate relief.

Dermochalasis, Blepharochalasis (Baggy or Puffy Eyelids)

Just as aging skin may appear loose and excessive elsewhere, so it can on the eyelids. When such excessive skin results in baggy, drooping lids it is called *dermochalasis.* It is usually of cosmetic significance only, causing the affected individual to look older and perhaps half-asleep. Very rarely, the skin is so redundant that it hangs down far enough to interfere with side vision. Simple procedures are available for removal of this excess skin, and they almost always make an individual look younger and more alert.

Baggy lids in younger individuals may be due to a defect in a fibrous tissue of the lid which holds the orbital fat back in place. Herniation of the fat into the space under the skin of the lid causes a similar puffy or baggy appearance. Again, simple procedures exist to restore a more normal appearance.

It is important that lid bagginess not be confused with true lid swelling (edema), which can accompany kidney and thyroid disease and other medical conditions.

Xanthelasma, Xanthoma (Yellow Spots)

Yellowish deposits under the skin of the lids are called *xanthelasma.* They are of medical importance because they may indicate an unsuspected underlying problem with the body fats, or lipids; in fact, the yellow plaques are actually cholesterol and lipid deposits themselves. While many patients with such spots are normal, a lipid profile—an analysis of the fats in the blood—along with a general physical examination should be performed as a means of detecting and preventing premature heart and vascular disease associated with elevated lipid levels. There are several methods for removal of unsightly plaques, but these have only cosmetic significance if blood and medical evaluations are normal. Removal under local anesthesia is easy if the deposits are not too extensive.

Conjunctivitis (Pink Eye)

Conjunctivitis is the medical term for the condition in which the conjunctiva, or outer lining of the eye and inner lining of the lid (figure 4–4), becomes inflamed due to allergy, irritation, or infection. Irritants such as smoke, smog, chemicals, and fumes can cause inflammation, as can foreign bodies such as insects and cinders. True allergy to such things as airborne pollens and animal fur also can result in "pink eye." Viruses, bacteria, and fungi are some of the micro organisms that attack the conjunctiva. The ophthalmologist has to distinguish carefully among the many causes of external inflammation and treat each cause appropriately. Not to be forgotten, of course, is the fact that other ocular problems such as acute glaucoma and intraocular inflammations can sometimes mimic simple conjunctivitis.

Corneal Dystrophies and Degenerations (Cloudy Cornea)

The cornea (see chapter 3 and figure 4–3) is the window of the eye and the component that contributes most to its focusing power. It is more than an inert lens, though: it provides us with crisp,

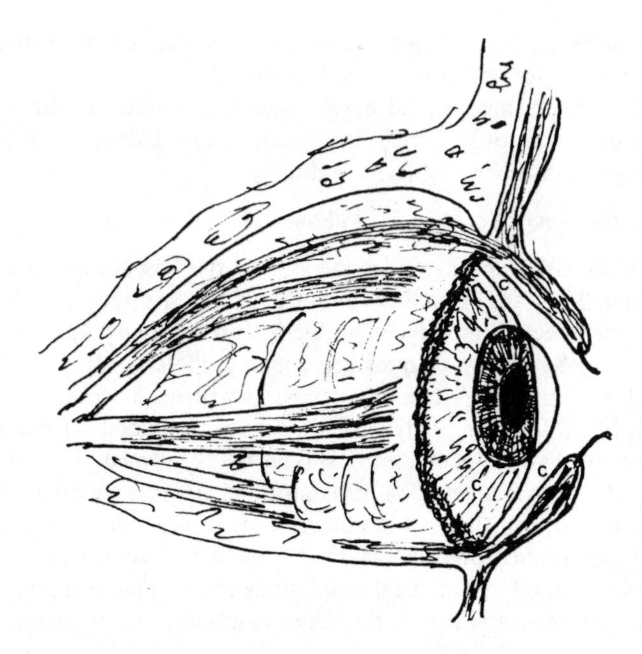

Fig. 4-4. The conjunctiva (c) is the thin, transparent mucous membrane lining of the inner aspect of the eyelids and front surface of the eyeball. The conjunctiva extends from the cornea into blind folds (cul-de-sac) above and below and then onto the inner lid surface. Thus foreign bodies cannot get out of the conjunctival sac and into the orbit without penetrating trauma.

clear images. A very active front layer of cells called the epithelium and an equally active and important posterior (inner) layer called the endothelium maintain the cornea's spherical curvature, clarity, and critical state of hydration (water content). These layers of cells pump excess water and wastes out of the cornea and ingest nutrients. There are a number of conditions in which one of these layers, or the intervening corneal stroma (the central bulk of the tissue), is imperfect. Called *corneal dystrophies* or *corneal degenerations,* these conditions are sometimes present at birth (often hereditary), though most develop later in life. If they result in a sufficient irregularity of the cornea or a disturbance in its clarity, a contact lens or a corneal graft operation may be necessary. These treatments will be described later.

Keratoconus (Conical Cornea)

One common example of a corneal dystrophy is *keratoconus,* a condition in which a weakness of the cornea causes it to become progressively more irregular and conical, almost pointed sometimes. Naturally, this distortion in the main focusing lens of the eye causes equal distortion in vision. When not too severe, a contact lens may be worn to provide a new front spherical focusing surface. If the cornea bulges too much, however, and a contact lens is not tolerated, a corneal transplant may then be necessary.

Endothelial Dystrophy

Another fairly common condition is one in which the endothelium, or inner lining, weakens later in life. The cornea can no longer maintain its normal state of hydration and excess fluid seeps into it. The cornea then thickens and takes on the appearance of ground glass. The vision then becomes clouded—as if one were looking through a steamed glass, with halos around lights, distortion, and general blurring.

Other conditions in which opacities form in the cornea cause varying degrees of distortion and blurring, depending on the location and extent of the opacities. Medications sometimes relieve corneal swelling and opacities; contact lenses may help to restore some vision. The new soft lenses can alleviate the pain and irritation that come with epithelial, or surface, breakdown.

Abrasions, Recurrent Erosions

Surface scratches or abrasions (epithelial loss) can be very painful thanks to the extremely rich nerve supply of the cornea. This nerve supply serves us well as part of the reflex mechanism that initiates blinking and tearing to avoid injury and damage. Its efficiency is apparent in the excruciating pain and tearing one experiences when the cornea gets scratched or when a foreign body such as a cinder or metal fleck lands on it. Fortunately, the epithelium is an active tissue which can spread over a superficially abraded area with complete healing and no scar formation in twenty-four to thirty-six hours. Intervening infection or persistent irritation can impede this process, so the ophthalmologist often treats such injuries with appropriate medicines and sometimes a firm patch to promote quick, uncomplicated re-epithelialization.

Rarely will an abrasion, or scratch, be deep enough to disrupt the bottom of the surface layer of epithelium. When this "basement membrane" is injured, healing may not be complete for several months or more. The loose new epithelium over the defect may then detach for seemingly little or no reason, and the patient experiences sudden pain, tearing, and redness. This problem is called *recurrent erosion*. It often occurs in the morning or late at night, when the moist conjunctiva of the inner surface of the lid may have dried somewhat, becoming slightly adherent to the cornea it is covering. With motion or opening of the lids, the loosely attached patch of epithelium pulls off. Ointments applied at bedtime will disrupt this cycle, as will a patch or a therapeutic soft contact lens.

Corneal Scars

Injuries that penetrate deeper than the superficial epithelial layer of the cornea will form a scar. The extent and location of the scar will determine whether or not it affects the vision. Fortunately, only the central few millimeters of the cornea, normally 12 millimeters (about ½ inch) in total diameter, constitute the optical zone through which we see. If vision is significantly disturbed and cannot be corrected with spectacles or a contact lens, corneal transplantation may be necessary.

Corneal Ulcers

A whole range of microorganisms (fungi, bacteria, viruses, parasites) can infect the cornea and cause ulcers and abscesses. When moderately deep, these ulcers can cause permanent scarring; still deeper ulcers can spread the infection inside the eye itself, a condition known as *endophthalmitis*. With infection inside the eye, damage to the ocular contents can cause blindness. A corneal ulcer thus deserves immediate, careful treatment. It may appear to the patient as a simple red eye, but most often it is associated with pain, sensitivity to light, watery or thick discharge, and variably blurred vision, depending on the location. The patient may even see a white spot on the cornea, appearing as a white spot over the iris (the colored part of the eye) or over the central black pupil. The cornea, you will recall, is the clear crystal dome that covers the iris and pupil.

Fungal ulcers are more common in warm, subtropical climes, but suspicion is always warranted after an eye injury from vegetable material such as a twig or an object contaminated with earth. These ulcers require special care and medication.

Bacterial ulcers, most commonly due to *Staphylococcus,* or "Staph," are usually amenable to antibiotics presently available. Diagnosis and tests for antibiotic sensitivity must be performed quickly, for the ulcer can spread quite rapidly, depending upon the specific organism involved.

Herpes Infections

Two viral ulcers deserve special mention. *Herpes simplex,* associated with cold sores, is especially dangerous when it infects the eye. Usually it strikes only the mucous membranes of the mouth or lip, but it may occur almost anywhere. If one allows such a sore, or the discharge from it, to get close to the eye, direct inoculation (i.e., direct introduction and spread of infection) may occur. Before this problem was understood, young children often contracted a Herpes ulcer of the eye after being kissed about the eye by an unsuspecting relative or friend with a cold sore of the lip. Herpes simplex corneal ulcers still affect about three hundred thousand Americans each year. Once the ulcer establishes itself, it is likely to recur at the same spot, just as cold sores tend to recur. Scarring and loss of vision are possible long-term results. Although new drugs are being developed to fight them, Herpes corneal ulcers remain a painful and disabling problem.

Despite the name, *Herpes zoster,* the virus causing shingles and chicken pox, is not related to Herpes simplex, although it, too, causes serious problems when it involves the eye. Shingles is a peculiar disease caused by a virus infecting sensory nerve roots near the spinal column or brain stem. Pain and then skin blisters and crusted sores develop along the distribution of the involved nerve. It is thus characteristically confined to one side of the body and one segment. If it involves the first division of the trigeminal nerve, the sensory nerve of the forehead and upper face, the eye is at risk. Serious corneal and intraocular inflammation can result, so the eye must receive immediate attention and evaluation. While the eyelid may be grossly affected, the eye itself may be spared. Conversely, minimal lid and face involvement can be associated with significant eye infections.

While most other viruses (with the exception of Herpes simplex) are not susceptible to any presently available medications, the viral infections themselves will usually run their course and subside when neutralized by the patient's own immune system. They leave little or no permanent damage unless bacterial infection has supervened or severe inflammation has occurred. The ophthalmologist watches for these complications and averts them with appropriate medications.

Corneal Transplantation

The damaged cornea must be replaced in order to restore vision after severe, irreversible corneal disease or scarring. To date, no artificial material is tolerated as well as other human tissue. Corneal transplants were among the first transplantation operations performed, and they are among the most successful. Because the cornea normally has no blood vessels, the transplanted tissue is not immediately identified as foreign and rejected by the body. When severe disease or injury has induced abnormal blood vessel growth in the patient's cornea, the chance of a successful corneal graft is reduced. Nonetheless, new surgical techniques and medications have greatly improved the prognosis even in such cases.

In most corneal transplant operations, the diseased or opaque central portion of the cornea is removed with a trephine, a round knife. A corneal "button" of the same size is then cut from the donor eye and meticulously sutured in place (figure 4–5). Impressive advances have been made in the development of nonirritating, biologically nonreactive microsutures. These new techniques cause little postoperative pain or discomfort, and hospitalization for an uncomplicated graft is usually no more than several days. Either general or local anesthesia can be used for corneal transplantation, depending upon the severity of the case.

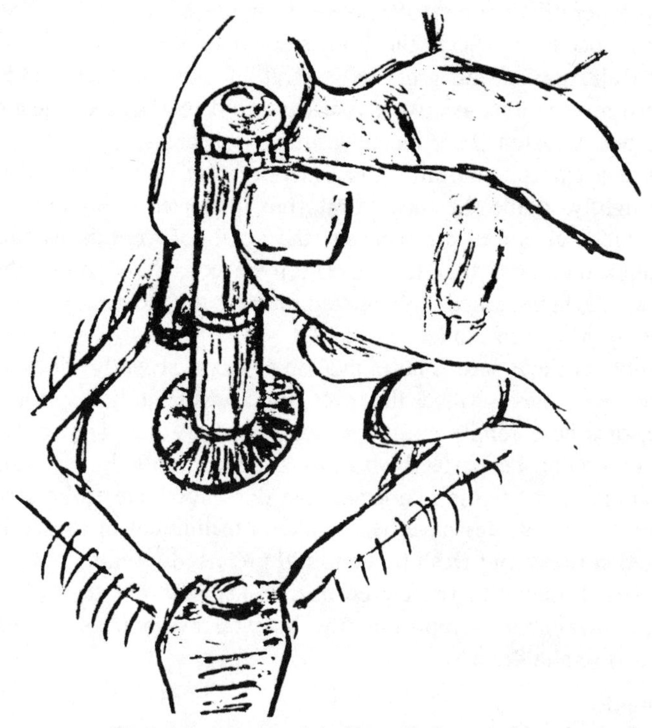

Fig. 4-5. A round knife called a trephine is used to remove a button of cornea in preparation for a corneal transplant. A button of identical size will be removed from a donor and meticulously sutured in place.

Complications from corneal transplantation are usually proportional to the severity of the disease for which the operation is performed. Astigmatism, or irregularity of the graft surface, may necessitate wearing a contact lens for the best clear vision. Tiny leaks along the graft margins may require resuturing, and infection can flare up. In rare cases, the manipulation necessitated by the operation itself may induce a cataract or aggravate an existing one. Occasionally, mechanical factors such as damage to the donor material, or irregular margins and a poor apposition of host and graft, may lead to clouding of the grafted cornea. Pre-existing or induced uncontrolled glaucoma may compromise the donor endothelium and cause graft swelling and clouding. Somewhat later in the postoperative course (after a week or more) a rejection reaction can occur in which the body's immune system attacks the graft. Rejection of the graft will result in clouding and failure. Modern medications are usually successful in reversing a rejection crisis, but occasionally a repeat grafting is necessary.

Corneal transplantation is thus an operation that is not undertaken lightly. While the success rate has improved strikingly with newer techniques and medications, the potential for serious complications mandates that the operation be considered only when alternatives have failed and a commitment to indefinite careful follow-up has been made.

Donor corneas are made available from recently deceased persons who have willed the eyes for transplantation prior to death, or whose family grants permission afterward. The corneas must of course be perfectly healthy to be acceptable for transplantation. If the tissue is normal, age does not seem to be a significant factor. Strides have been made in techniques of preserving donated corneas, but fresh tissue is still preferred by most ophthalmologists. Patients to receive corneal transplants are thus placed on call—available to report to the hospital as soon as a donor cornea is available.

Eye Banks

Eye banks have been established all over the United States in order to coordinate the availability of tissue for transplantation. The "ham" radio network is used to broadcast to eye banks across the country the availability or the emergency need of tissue. Airlines then speed the donated corneas to their destination. Nonemergency patients are placed on a waiting list to be summoned when a donor cornea is available. Those interested in willing their corneas for transplantation can do so by contacting a local eye bank.

Tears and the Lacrimal System

As we discussed briefly in chapter 3, the eye and its external tissues require constant moisture to maintain smoothness and, in the case of the cornea, transparency as well. The tears serve this lubricating function in addition to carrying nutrition and antibodies to ward off infection. The *tear film* is actually a three-part, sandwich-like coating of the eye. The outermost layer is an oily film derived mostly from the Meibomian glands of the lids. The intermediate watery layer is derived from the large tear gland in the upper outer corner of the eye socket (figure 3–8) and from smaller glands at the ends of the upper and lower folds of the conjunctiva. The innermost layer is mucous, a substance produced by special cells of the conjunctiva. The tears circulate constantly, draining to the inner corner of the eye, where an upper and lower punctum collect the fluid. It drains through tiny canaliculi to the tear sac, then down into the back of the nose to the throat. This drainage system is the reason why one can often taste drops put in the eyes.

The blinking action of the lid muscles creates a pumping effect on the tear sac. With lax, loose lids, as in ectropion or a Bell's palsy, the pumping mechanism is faulty and the tears fail to drain properly. Stagnation of the tears breeds infection, as does a tear deficiency.

Keratoconjunctivitis Sicca (Dry Eye)

Keratoconjunctivitis sicca can set in after scarring conjunctival inflammations have destroyed the mucous cells, tear glands, and tear ducts or after certain problems with the tear glands. Tearing deficiencies are usually of unknown origin, related to aging and hormonal state and most common in postmenopausal women. The salivary glands of the mouth may also be deficient, causing a dry mouth. Systemic disease processes such as rheumatoid arthritis may be associated with tear deficiencies. For whatever cause, tear deficiency is at best occasionally annoying and uncomfortable, and at worst disastrous. Redness, inflammation, and a gritty, sandy feeling, as if something is in the eye, are the usual symptoms of "dry eye." The reason for this inflammation and irritation is that the epithelium of the conjunctiva and cornea—their outermost layers of cells—dry and sometimes slough. This disturbance of the surface results in inflammation and the sensation of foreign material in the eye.

A concomitant of decreased tearing is a deficiency of the antibodies normally contained in the tears. Chronic inflammation and inadequate cleansing promote infection. In the most severe cases, with virtually no tears produced, the normally moist transparent conjunctiva and the beautifully crystalline cornea become dried. The surface layers of cells become like skin cells, which regularly produce dried debris known as keratin. The result, then, is a rough and opaque coating of the eye, both chronically irritating and progressively crippling to vision.

In mild to moderate cases of dry eye, simple artificial teardrops will suffice. Patients should either avoid conditions that will aggravate the dryness or be prepared to adapt to it. Wraparound glasses or side shields with glasses should be worn on very windy days. Hair dryers and blowers should be avoided, and artificial teardrops or lubricants should be brought along on trips to very dry climates. When simple drops and ointments prove inadequate, the ophthalmologist may want to block off the tear drainage channels, the puncta (figure 3–8), in an effort to prevent the remaining tears from draining away rapidly. The new soft contact lenses provide some relief and protection, but they also are subject to drying and are not always suitable. Doctors have been testing the use of a reservoir of artificial tears bounded by membranes that control the release of the solution. Such thin, waferlike discs are presently used to deliver low, constant, round-the-clock dosages of medication to glaucoma patients.

Epiphora (*Tearing, Wet Eye, Blocked Tear Duct*)

At the other end of the spectrum is the eye that waters too much. The lid skin may become macerated, thickened, and red from constant wetting and wiping. First to be ruled out in all such cases is a source of chronic irritation: a foreign body hidden up in the superior cul-de-sac or under the upper lid, chronic allergy, or lid inflammations. If no such problem exists, then the tear drainage mechanism may be faulty.

In newborns the obstruction to tear flow almost invariably rests at the nasal end of the nasolacrimal duct (figure 3–8). If the duct has not completely developed, a thin membrane will persist over the opening in the nose. It may open spontaneously, but if it does not, it can be forced open by injecting fluid through the puncta and canaliculi, or by probing gently with a thin, flexible, blunt-tipped wire.

Dacryocystitis (Infected Tear Sac)

An infection in the tear sac, the common cause of blocked tear duct after infancy, can lead to scarring and permanent obstruction to tear drainage at the end of the sac. The obstruction may be due either to a mild unnoticed infection or to one or more severe acute episodes. Such infections require vigorous therapy with appropriate antibiotics.

If the obstruction is at the junction of the canaliculi and the sac (called the common canaliculus), it will bring on tearing and bouts of conjunctivitis. But the tear sac will not swell up at the innermost corner of the lids. Pressing over this area (massaging a swollen tear sac against the nose) pushes the tears, mucous, and pus back through the puncta into the eye when the obstruction is at the lower end of the sac. Following the course of dye-tinged fluid, the ophthalmologist can localize the site of obstruction. If simple irrigation or probing does not help, the doctor can perform surgery to create a new pathway for the tears to drain into the nose. Although the operation enjoys a high rate of success, each case must be considered individually, for ancillary conditions can alter the prognosis and advisability of this procedure.

Eye Make-Up

Eye cosmetics are widely used and often abused. Although they are usually attractive and appealing, they can create many needless problems. Actual allergy to the cosmetics is probably far less common than simple irritation or infection from poorly applied cosmetics or contaminated materials. The ophthalmologist can usually distinguish the various causes of inflammation related to cosmetic use. Occasionally, more than one factor is involved, such as allergy and infection together.

One should understand the anatomy of the eyelid before applying make-up. The pink rim of the lid behind the row of lashes contains a row of special oil glands that contribute to the tear film. If one applies eye liner along this rim, or mascara at the base of the lashes where it may get on the edge of the lid, some of the openings to these glands may get blocked. Once blocked, the glands are prone to infections and styes. Microscopic particles in the make-up sometimes lodge in the conjunctival sac behind the lid, where they act like a foreign body that causes inflammation, if not abrasions. The particles of make-up may become embedded under the normally smooth conjunctiva. In addition to blocking the small oil glands along the lid margin, make-up, especially mascara, can also block the hair follicles of the lashes. These tiny root canals at the base of each lash are prone to infection if they become blocked.

Thus, whether one uses hypoallergenic (less likely to cause allergy) make-up or not, the key is proper application. Eye liner, shadow, and the like should always be applied above the lashes of the upper lid and below the lower eyelashes. In this way no make-up should get near the edges of the lids or into the lash follicles. Mascara should be placed along the outer tips of the lashes, but not down to the base of the lashes lest it occlude the lash follicles (figures 4–6 and 4–7).

If an eye infection develops, one should immediately consult an ophthalmologist and discard all used eye cosmetics. Eye infections may be transmitted back to the container of used make-up on the applicator, and any further use of the make-up will probably reintroduce the infection into the eye or lids. Microorganisms can persist in a closed container for an indefinite period of time.

If eye or lid inflammation persists despite these precautions, allergy or an underlying skin condition should be considered. A different brand of cosmetics or a hypoallergenic variety may be tried, but some rare individuals cannot wear any cosmetics without difficulty.

Fig. 4-6. Mascara should be applied to the outer tips of the lashes only, not to the bases or the pink eyelid margin.

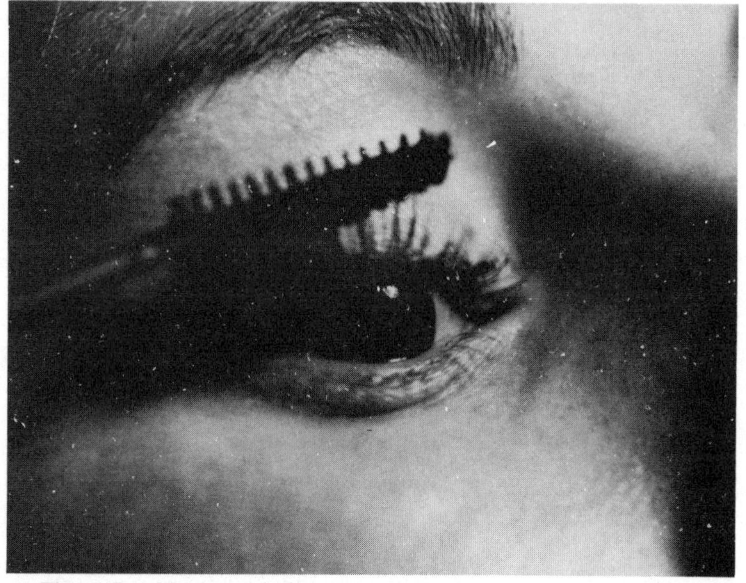

Fig. 4-7. Mascara similarly applied to the outer portions of the upper eyelashes.

5 Glaucoma

Glaucoma is the name used by ophthalmologists to describe those conditions in which the pressure within the eye is abnormally elevated, thereby damaging the optic nerve. There are several types of glaucoma, but common to them all is the problem of increased intraocular pressure and its effect on the retina and optic nerve. The exact mechanism by which this pressure damages the optic nerve is still under investigation. Increased pressure may impede nourishment of the nerve at its entrance to the eye, either by compressing tiny blood vessels that supply the nerve head at that point or by impeding the flow of nutrients along the nerve fibers themselves.

Evidence of such damage to the optic nerve presents itself as progressive defects in the peripheral vision. With the aid of an ophthalmoscope the eye doctor can see parallel progressive atrophy and excavation of the optic nerve.

Visual field is the term used to describe the extent of one's total vision from side to side, centrally as well as peripherally. There are various methods of examining the visual field. In the simple *confrontation technique,* an examiner moves his finger or an object off to the side while the subject stares straight ahead. The subject responds when he sees motion or the object off to the side. At the opposite extreme, ultrasophisticated devices exist to control background illumination, contrast, intensity, and size of test objects.

The *tangent screen,* a simple flat surface upon which little beads of various sizes and colors are presented, is the simplest test for the central field and the one most commonly used for glaucoma (figure 5–1). Various types of perimeters are available to test peripheral as well as central vision (figure 5–2). The important factor is really the constancy of the testing situation, for progressive change in the field is what determines the urgency and adequacy of the therapy once a diagnosis is made.

Under ideal circumstances, the visual field is a critical measure of the status of the optic nerve and its response to glaucoma; many factors, however, reduce the accuracy and sensitivity of the testing. Though an absolute necessity, patient cooperation is often difficult to obtain. The eye must be fixed on a given point straight ahead, but attention must be directed to the periphery so that one can respond as quickly as possible when the test object (a light or small round bead) appears. The eye's tendency to look in the direction of an approaching object must be fought actively. To the extent that fixation varies and shifts, one is no longer accurately determining the vision in the periphery. Ancillary problems that reduce vision not only make fixation more difficult, but also cause visual field changes of their own. Cataract, macular degeneration, and retinal scars are concurrent conditions that make evaluation of the visual fields for glaucoma more difficult.

If visual field cannot be assessed, the ophthalmologist must then rely more on the appearance of the optic nerve head to estimate the damage done to vision by glaucoma. The parallel reflection of this nerve damage is loss of the visual field. As the damage progresses, the nerve head appears progressively more pale and excavated, or cupped, and the retinal vessels emerging from it will be shifted to the side. The ophthalmologist makes notes on the extent of cupping, pallor, and shift of vessels in order to detect any changes. He may draw the disc (as the optic nerve head in the eye is called) or take pictures of it for this same purpose.

The optic nerve head, described in chapter 3, is the one point in the back of the eye where there are no light-sensitive retinal cells. It thus appears normally as an absolute blind spot in our visual field. It is not normal to have any other such blind spots, called *scotomas.*

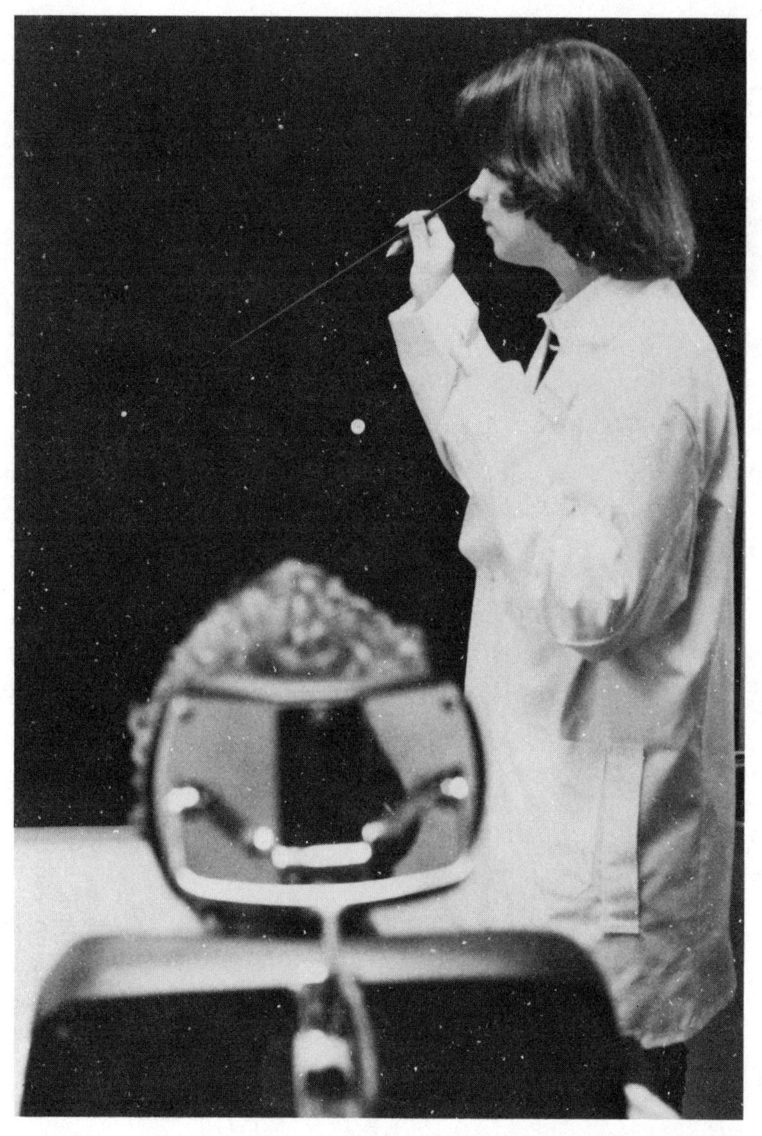

Fig. 5-1. The central visual field is examined on the tangent screen. With one eye occluded, the other eye maintains fixation on the central spot while a test object is presented in various positions over the field.

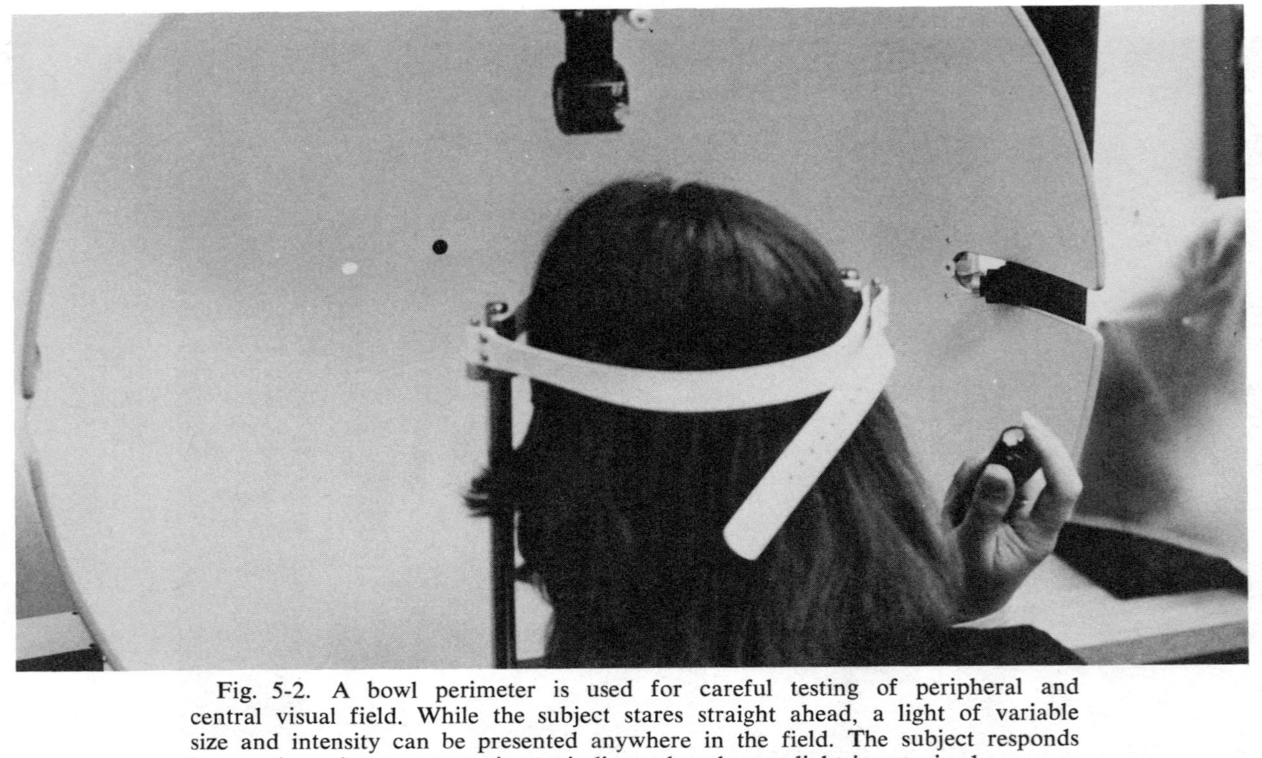

Fig. 5-2. A bowl perimeter is used for careful testing of peripheral and central visual field. While the subject stares straight ahead, a light of variable size and intensity can be presented anywhere in the field. The subject responds by pressing a button or tapping to indicate that the test light is perceived.

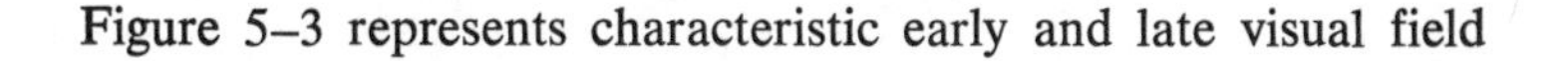

Figure 5–3 represents characteristic early and late visual field

Fig. 5-3. Visual field plots indicating progressive loss of field characteristic of uncontrolled glaucoma. The glaucoma patient sees only in the restricted white areas.

defects in glaucoma, and figure 5–4 depicts an optic nerve with varying degrees of glaucomatous damage. Note that the early defects are very subtle; in fact, the normal absolute blind spot is not apparent unless one looks carefully for it, as demonstrated in figure 3–4. Thus, by the time a patient notices visual impairment, there has been extensive and usually irreversible damage to the nerve (figure 5–5). Ophthalmologists strongly recommend routine checkups approximately every two years after age forty when glaucoma occurs most frequently. An eye doctor can determine quickly and easily whether the pressure in the eye is normal, suspicious, or abnormally high.

Tonometers measure the pressure in the eye. The Schiotz tonometer (the most widely available) is placed on the eye after a drop of numbing medicine is applied. The indentation it creates in the cornea is measured and translated into a pressure reading (figure 2–7). Another instrument used more often in the ophthalmologist's office is the applanation tonometer. It has a calibrated spring balance from which a given force is applied to a small plastic disc. The disc flattens a tiny area of the cornea, and the force necessary

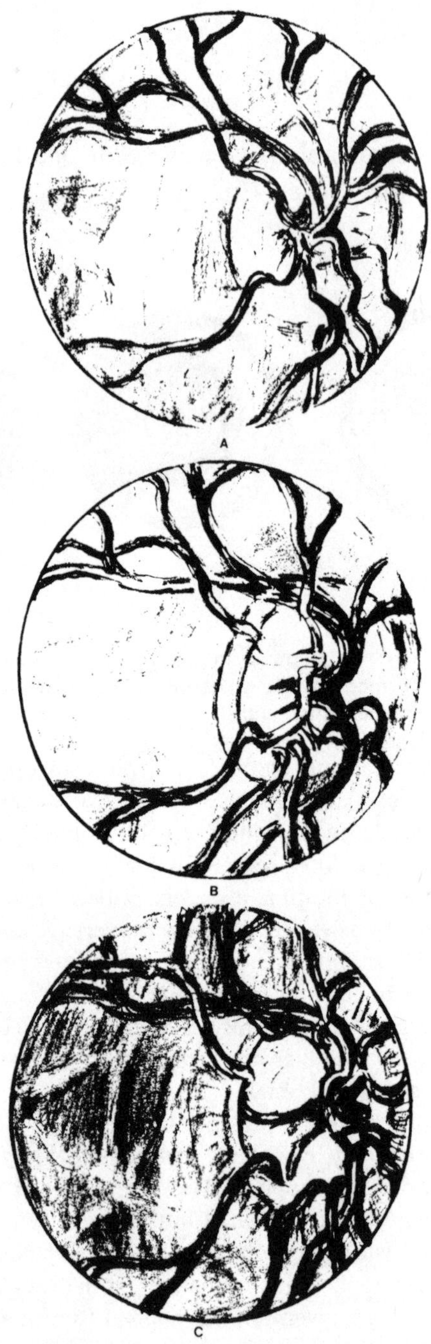

Fig. 5-4. Representations of the progressive excavation of the optic nerve head in the eye due to glaucoma. The stages A through C of cupping and atrophy correspond roughly to the stages of visual field loss A through C noted in fig. 5-3.

a

Fig. 5-5. A landscape (a) as seen by a patient with glaucoma advanced to a stage at which the limited field will be noticed by the patient and will represent a disability (b).

b

for this minimal flattening is then read directly as the pressure (figure 5–6).

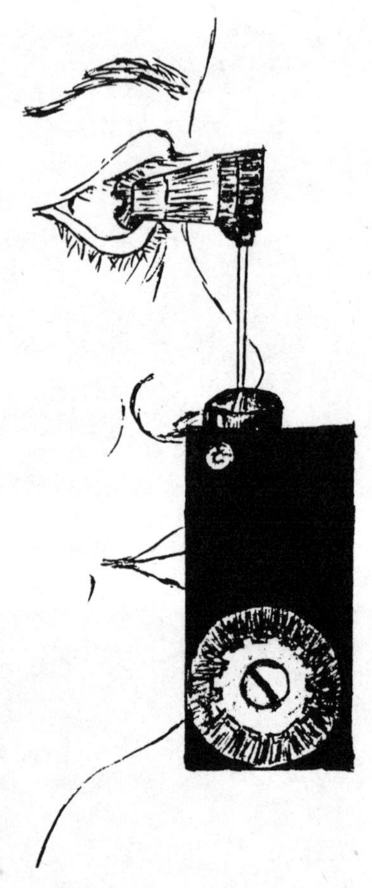

Fig. 5-6. An applanation tonometer, most commonly used as an attachment to the slit lamp (see fig. 2-8.).

The pressure within the eye is expressed scientifically as millimeters of mercury, i.e., the amount of pressure able to support a column of mercury at that height. Pressure is normally in the teens. When it reaches the low twenties it becomes suspicious. The higher the pressure beyond that level, the greater the certainty of glaucomatous damage to the nerve. As always, though, variables must be considered.

In general, the younger the patient, the better the optic nerve will resist potentially damaging pressure. The higher the blood pressure, the better the nerve seems to withstand pressure within the eye. Though one should not avoid therapy for high blood pressure because of glaucoma, if a glaucoma patient's blood pressure is reduced significantly, the ophthalmologist must be advised of the possible need to alter glaucoma therapy.

The pressures in the eye vary from time to time in the normal person, just as blood pressure and other physical measurements vary. Normal always represents a range, and in fact represents a range peculiar to each individual. An unequivocal diagnosis of glaucoma rests on the demonstration that the net effect of the given range of pressures in that person is damaging to the optic nerve.

Glaucoma "Suspects"; Ocular Hypertension

Pressure that is damaging and abnormal in one person may be entirely normal in another; hence, many people who exhibit ocular hypertension show no sign of glaucomatous damage from this suspicious pressure. Most ophthalmologists agree that these glaucoma suspects should be monitored rather than subjected to possibly unnecessary therapy. With careful examinations for early field loss or changes in the appearance of the optic nerve, the ophthalmologist will detect the first sign of glaucoma long before it represents any significant nerve damage, and certainly well before visual loss is apparent to the individual. One large, well-documented study showed that of thousands of ocular hypertensives monitored for a long period of time, only a very small percentage actually developed glaucoma. They can thus be reassured that they are unlikely to develop glaucoma and that with careful periodic evaluations they will not develop any significant impairments even if a diagnosis of glaucoma is made later. On the other hand, these individuals should adhere strictly to their ophthalmologist's advice. Playing the odds and neglecting periodic checkups is very risky when one considers the stakes—severe visual disability leading to total blindness. Vision usually fades slowly, so one is not aware of the condition until it is very advanced.

Glaucoma Screening and Tests

It is estimated that at least two people out of one hundred over age forty have suspicious intraocular pressures. That translates into several million individuals in the United States alone. Fortunately, routine screening tests (tonometry) are now harmless, and quick. Thanks is due public service groups like the Lions Clubs, who organize mass screening clinics to find potential glaucoma victims. Bear in mind, however, that these screening tests, like the vision tests administered to children in school, are designed to miss as few people with real problems as possible. Thus, many people who are only suspect or who, on retesting, have no problem at all, will be advised to seek further evaluation. It is impossible to be more discriminating in screening evaluations without missing larger numbers of truly affected individuals. As long as people understand the preliminary nature of these exams and do not get overly anxious about the results, no harm can ensue.

Symptoms of Glaucoma

As mentioned initially, glaucoma is the name for many conditions which have in common abnormally elevated pressure within the eye. The previous discussion is most appropriate to the common type of glaucoma associated with aging. Called *chronic simple,* or *open-angle,* glaucoma, this type progresses slowly and insidiously with virtually no symptoms until 90 percent or more of the optic nerve has been destroyed. Vision may remain a perfect 20/20, but only the central 10 percent or less of the total visual field remains. Any of the glaucomas in which pressure does not rise suddenly or to exceedingly high levels can follow this insidious course.

A sudden increase in pressure or progressive elevation to a high enough level will damage the inner lining of the cornea and allow fluid to seep into its inner layer. Instead of looking through a crystal clear cornea the subject is now looking through a swollen cornea similar to ground glass. At this point, definite symptoms develop. One may have pain and a dusky redness in the eye. Vision becomes blurred because light is being diffracted into halos. A halo effect appears around lights at night. The pain and generalized stimulation of the nervous system may be great enough to cause such systemic symptoms as nausea and vomiting. Patients have been known to go to their family doctor complaining primarily of nausea, vomiting, and headache, dismissing the blurred, red eye as incidental. In fact, severe glaucoma was behind all the symptoms.

Intraocular Pressure and the Causes of Glaucoma

How, then, does this excessive pressure develop in the eye? In chapter 3 we mentioned that the inner structures of the eye (the lens and inner cornea) are nourished by a clear fluid called the aqueous humor. (If these structures contained blood and blood vessels, they would not be crystal clear and we could not see.) The fluid is produced by the ciliary body, a structure behind the iris that contains muscles to tighten and loosen the suspensory fibers (zonules) of the lens in focusing. From the ciliary body the aqueous flows around the lens and through the pupil into the anterior chamber. From the anterior chamber it exits via the trabecular meshwork in the filtration angle, the 360-degree angle formed by the cornea and iris. The trabecular meshwork is composed of layers of sievelike plates through which the aqueous filters into the microscopic circumferential Canal of Schlemm (figure 3–3). From here the aqueous empties back into the bloodstream via the veins that drain the eye. Glaucoma impairs the exit of aqueous humor, causing an elevated pressure within the eye. The nature and site of impairment determine the type of glaucoma.

Open-Angle, or Chronic Simple, Glaucoma

The filtration angle of the eye can be viewed through a special contact lens with a built-in angulated mirror. This quick and painless examination, called a *gonioscopy,* yields tremendously important information. If the angle appears normal and open, with approximately a 45-degree approach, the glaucoma is labeled *open-angle,* or *chronic simple*. The slow, progressive loss of peripheral field that characterizes this form of glaucoma is not apparent early without special field examination, and the pressure usually does not reach levels high enough to cause corneal swelling (resulting in blurred vision and halos around lights), redness, pain, and discomfort. Screening programs often detect this insidious and common type of glaucoma.

Acute, or Closed-Angle, Glaucoma

A narrow angle indicates to the ophthalmologist that the patient has or may develop *acute glaucoma*, so named because it usually presents as a sudden, severe increase in intraocular pressure associated with some or all of the symptoms previously mentioned. The reason for this sudden elevation of pressure and its attendant symptoms is that the narrow approach to the filtration angle has suddenly blocked off completely, allowing no exit of aqueous humor from the eye.

The normally deep angle of the eye is ample to accommodate the slight bowing forward of the iris which occurs when the pupil is about half-dilated and in a position to resist the normal flow of aqueous into the anterior chamber. This slight bowing forward pops open the space between the lens and iris, and the aqueous flow resumes unimpeded. In the narrow angle, this slight bowing forward of the iris may actually block the filtering meshwork, causing an immediate increase in anterior chamber pressure. More aqueous builds up posteriorly trying to pop open the iris from the lens, but in so doing it only pushes the iris up against the angle structure more firmly and the angle is then closed. A vicious cycle has been started, and the pressure can reach very high levels.

Under high pressures the main vein or artery in the retina may collapse, causing irreversible loss of vision. Such high intraocular pressure also may damage the iris, the lens, and even the blocked-off trabecular meshwork, which may not function well even after opening (figure 5–7).

The condition is first treated with eyedrops that move the pupil —usually constricting it—to break up the pupillary block, i.e., the inability of the aqueous to flow through the pupil. In some instances, the doctor will recommend oral or intravenous medications to lower the pressure in the eye. Definitive therapy consists of creating an alternative pathway for aqueous to flow from

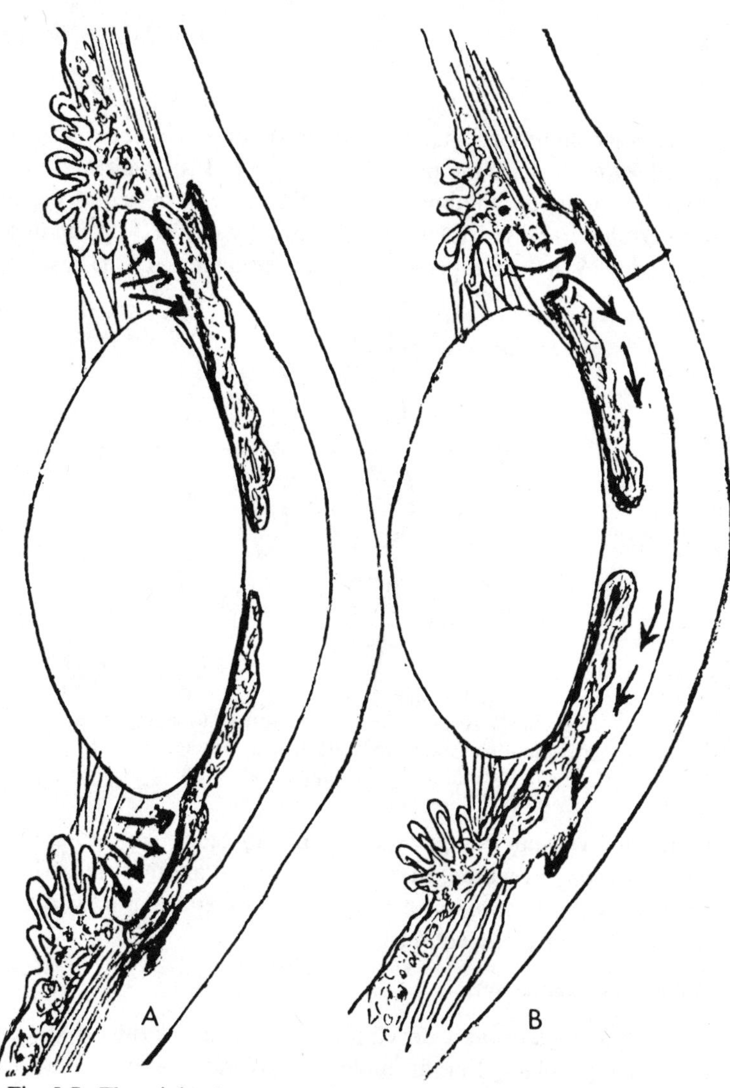

Fig. 5-7. The minimal normal resistance offered by the iris against the lens to aqueous humor flow through the pupil causes anterior bowing of the peripheral iris and release of the resistance. In eyes with a narrow filtration angle, this peripheral bowing can block the outflow channel and initiate a vicious cycle of increasing pressure in the eye that results in acute glaucoma (A). A peripheral iridectomy is simply a hole in the peripheral iris that provides an alternate route for aqueous to enter the anterior chamber and thus exit through the filtration angle. Pupillary block leading to acute glaucoma can no longer develop (B).

posterior to anterior chamber, circumventing the pupil. A *peripheral iridectomy,* as the procedure is called, is a simple and quick operation with few complications. A tiny hole is placed in the upper periphery of the iris, where it is usually covered by the upper lid (figure 5–8). It represents a cure in that once it has been

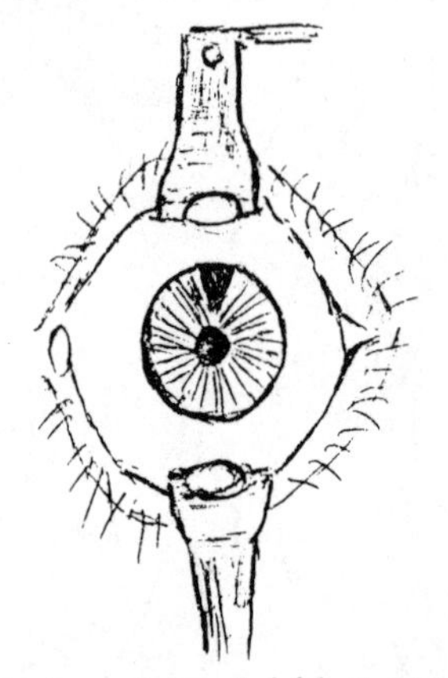

Fig. 5-8. A peripheral iridectomy appears as an inconspicuous black spot at the upper edge of the iris, usually hidden by the upper lid.

performed without complication, pupillary block can never recur. Eyedrops may break an attack and postpone another, but without iridectomy the possibility of pupillary block still exists even when drops are used conscientiously.

Secondary Glaucomas

Secondary glaucomas develop as a result of a primary systemic or ocular problem. For example, an overripe cataract may leak some of its protein through its capsule into the eye. This excites a foreign-body response in which large cells called *macrophages* invade and ingest the foreign material. The cells then attempt to flow out of the eye with the aqueous but, because of their size, wind up blocking the trabecular meshwork. The result is sudden and severe glaucoma. Similarly, extensive bleeding in the anterior chamber from trauma may block the outflow of aqueous.

Blunt trauma may actually tear part of the filtering angle, and if the angle is scarred it will not function properly. Inflammation within the anterior part of the eye may leave adhesions, or scars, on the peripheral iris, up against the filtering meshwork (see chapter 11). The outflow is thus obstructed, and glaucoma sets in. These scars are called *peripheral anterior synechiae.* When inflammation causes the pupil to scar down backwards to the lens, the scars are called *posterior synechiae.* They may deform the normally round shape of the pupil (figure 5–9). If they completely block the pupil, aqueous will be trapped in the posterior chamber and again severe glaucoma may result. An iridectomy must be performed in this situation to allow aqueous to flow into the anterior chamber by an alternate route. Diabetes and other vascular diseases of the eye may occasionally lead to glaucoma through the formation of tiny abnormal new blood vessels and scar tissue over the angle structures.

These examples are only a few of the many conditions that can impair the outflow of aqueous and thus induce secondary glaucoma. All totaled, however, they are few in number compared with the number of cases of primary, open-angle glaucoma.

Congenital Glaucoma

Another fortunately rare form of glaucoma is that appearing at birth or very early in infancy. *Congenital glaucoma* may be either an isolated, primary developmental anomaly or one of multiple genetic defects. Up to age three, the thick fibrous outermost coat of the eyeball (called the sclera) and the cornea are relatively elastic and distendable. Increased intraocular pressure before or shortly after birth, then, can stretch and enlarge the whole eyeball. The ancient Greeks used the word *buphthalmos,* meaning "ox eye," to describe this condition. With sufficiently elevated pressure the cornea will be swollen and cloudy as well.

Early symptoms in the young child include marked sensitivity to light, almost constant squinting and squeezing of the lids, and frequent tearing. Parents might mistakenly assume that the child is retarded because he buries his face in a pillow most of the day to avoid the discomfort of lighted surroundings. The condition is rare, but a child with these symptoms should be evaluated by an ophthalmologist.

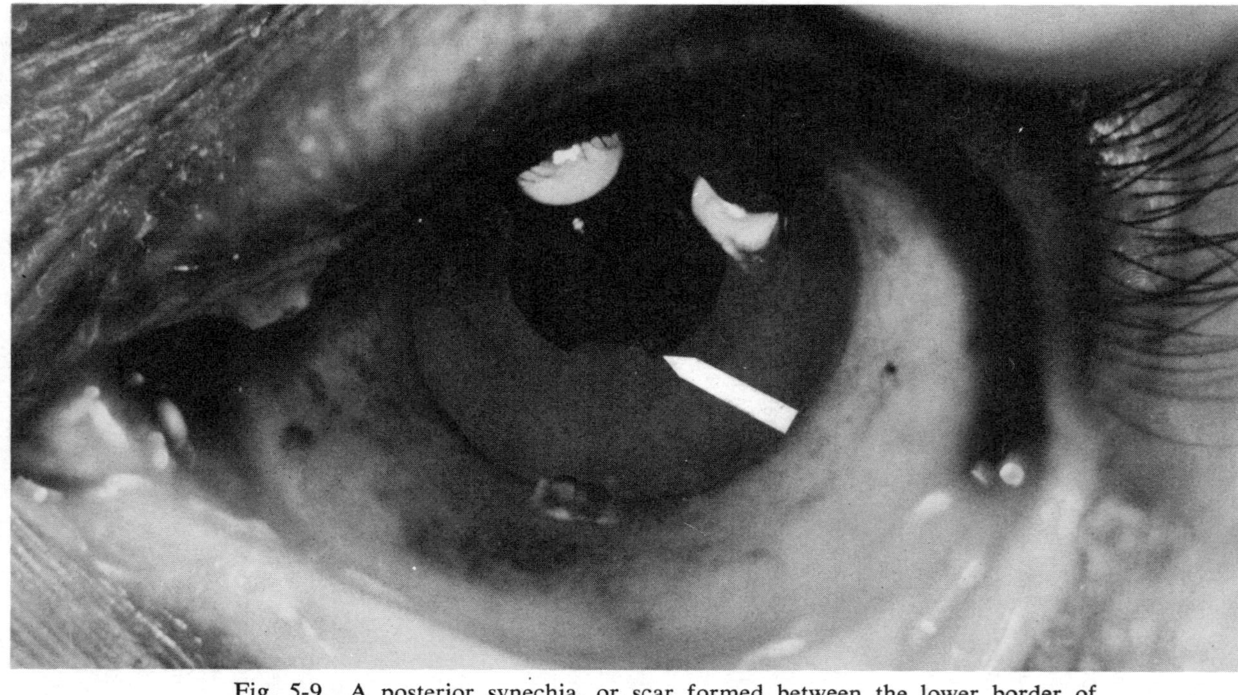

Fig. 5-9. A posterior synechia, or scar formed between the lower border of the pupil and the lens, deforms the lower edge of the normally round pupil. Such a scar is evidence of prior or present inflammation within the eye. It may be difficult to detect in an eye with a dark iris (arrow).

Glaucoma Treatment

Treatment of glaucoma varies with the type. Congenital glaucoma is usually treated surgically by opening up the trabecular meshwork. As discussed before, acute and secondary pupillary-block glaucoma are best treated with an iridectomy, which creates a new path for aqueous to flow into the anterior chamber (figure 5–7). The first line of therapy in open-angle and most secondary glaucomas is medical, however, operation being reserved as a last resort.

Tremendous medical and surgical advances have been made recently. Eyedrops are available to facilitate the outflow of aqueous from the eye and to suppress its formation as well. One of the common drops has as a side effect constriction of the pupil and stimulation of the ciliary muscle, the focusing muscle of the eye. This constriction can be most annoying at night, when the pupil would otherwise dilate to admit more light. Individuals with early cataracts who had previously been seeing well enough through large pupils which admitted enough light may find themselves visually disabled from the small pupil caused by these drops. Younger individuals whose lens still responds to the action of the ciliary muscle (see chapter 3) will find themselves transiently nearsighted from about one-half to an hour after the application of these drops, until about four hours later.

An exciting new method of applying these medicines is a tiny wafer inserted in the conjunctival sac. It is actually a reservoir of the medicine bounded by membranes that release small doses of it steadily throughout the day. It has to be replaced only once a week, so it is far more convenient than drops that must be applied several times a day. Since it releases the medication continuously, the wafer eliminates most of the side effects of sudden high dosages. These wafers are usually easy for the eye to adjust to, but they cost more than the drops.

If the drops are not tolerated due to allergy or side effects, or if they fail to lower the pressure, pills may be added to the treatment. Oral medicines reduce pressure by diminishing the production of aqueous humor. They happen to be mild diuretics as well and will stimulate urination for the first few days until the kidneys adapt. Fortunately, the eye does not so adapt, and the medicine's effect on intraocular pressure continues. There are certain side effects, most quite minimal, that some people experience with this, as with any, medication. It is not wise to list them since most are subjective, and suggestion often elicits or magnifies such problems. The ophthalmologist is very familiar with all the side effects and their significance. One should mention anything out of the ordinary and should, of course, advise the doctor of all other medications being taken, and of anything unusual about the diet, just as one should inform any other doctors of the eye medicines that were being used.

It must be stressed that these medicines all act for a given period of time only, after which the pressure in the eye will again begin to rise. Like blood pressure medicine, glaucoma medicine must be taken as directed by the eye doctor for an indefinite period of time. Glaucoma, like high blood pressure and diabetes, is a disease that is controlled by medicines, but not cured by them after a short period. But well over 90 percent of all glaucoma can be controlled with medications alone. Patients need not fear blindness, for they will lose no vision once under treatment.

When glaucoma is not controlled by medication, surgical intervention is necessary to prevent further irreversible loss of vision. New microsurgical techniques are quite promising in reducing the complications of surgery. These procedures can cure glaucoma by providing an alternate path for the aqueous to exit the eye. They are thus called filtering procedures because they establish a fistula, or hole, through which the aqueous humor can flow out of the anterior chamber and into the space beneath the conjunctiva. It is then absorbed directly into the bloodstream.

The success rate of these procedures as a whole is probably greater than 80 percent. The most common complication is a simple failure of the hole to stay open and thus control the pressure. But excess aqueous humor also may escape the eye if the fistula is too large; then the anterior chamber, the fluid-filled space between lens and cornea (figure 3–3), remains too shallow and the eye too soft. Cataract formation may be stimulated or aggravated, and in rare cases hemorrhage or infection may occur. Nonetheless, the vast majority of these procedures do work and succeed in controlling a disease that would otherwise lead inevitably to total blindness. These operations, like cataract procedures, are usually done under local anesthesia, involve relatively short operating time, and cause minimal or no discomfort afterward.

In summary, glaucoma is a disease that can be controlled in almost all patients who are willing to adhere to therapy and be monitored by their ophthalmologist. The prescribed medications may be considered a nuisance, but they are rarely worse than that—a small price to pay for vision. Not long ago, a diagnosis of glaucoma meant inevitable blindness. It is no longer so. We may not cure glaucoma with a short course of therapy, as we do pneumonia, but we can almost always control it and preserve sight indefinitely.

6
Cataract

Cataract is an extremely prevalent problem, occurring in 65 percent of patients over age fifty, and 95 percent of patients over age sixty-five. Over four hundred thousand operations to remove cataracts are performed each year in the United States alone, and almost two million Americans are known to suffer at least minimally from cataracts.

Although cataract is one of the most common surgical problems diagnosed, confusion and misconceptions about it remain surprisingly widespread. New treatment techniques, awarded considerable publicity in the lay press, have only added to the bewilderment of most cataract patients who have tried to educate themselves to the proper course of action. In fact, there is some disagreement in the ophthalmologic profession itself regarding the most recent innovations.

This chapter will clarify what a cataract is and explain how it can affect one's sight. In an effort to convey the current consensus of opinion in the ophthalmologic community, the chapter will look objectively at the problems and risks attending various treatment plans, indicating where and why controversy exists. In the end, the reader will have a basis from which to question the ophthalmologist about the alternatives available. The treatment of cataract is remarkably safe and effective. While the most recent innovations may be controversial and thus perplexing to the patient, it is a credit to the profession that complacency with the already safe, established procedures did not develop. Selecting among the alternatives should be a welcome opportunity to find the best of several possible treatment plans, rather than a desperate exercise in avoiding what is wrong or bad.

What, then, is a cataract? Does it have to be removed as soon as it is detected? Will it cause complete blindness if not treated immediately, or must we wait until it ripens? Can anything be done to halt or reverse the progression of a cataract? Do special diets or vitamins help? Is the eye removed during the operation and then put back? These questions and many others are constantly put to the ophthalmologist by patients who want to understand their problem and the proposed management of it. Hopefully this chapter will answer them all.

What Is a Cataract and How Does It Affect Vision?

Though incorrect, the idea that a cataract is a "skim," or "film," over the eye or its lens has definite historical precedent. About two thousand years ago, when the first descriptions of cataract were recorded, the Greek and Western cultures considered it to be a dried-up substance sitting in front of the lens in the eye. Only two hundred years ago French surgeons demonstrated that a cataract is in fact the lens itself which has become cloudy or opaque. This idea was professed earlier in Eastern literature, and a form of treatment that probably dates to prehistoric times was prevalent there as well. This technique is called *couching,* and it consists of poking needles or thin knives into the eye to push the cataractous lens down into the posterior cavity of the eye (figure 3–2). This technique is followed by an extremely high rate of complication which usually entails loss of the eye. Nonetheless, it is still practiced in some underdeveloped areas of the world.

When the cataract, which we now understand is an abnormal clouding or opacity of the lens, is removed from the pupil, or *visual axis* of the eye, the light rays can then pass unobstructed to the back of the eye, where they hit the retina, excite the receptor cells, and cause an impulse in the optic nerve.

Chapter 3 compared the eye to a camera. The retina is the film: it takes the light rays and establishes the patterns of light, dark, and color. The pattern is then transmitted via the optic nerve to the brain, where we actually see. The cornea, or crystal, focuses light as it enters the eye, while the lens of the eye adjusts the focus of the light rays that are allowed to enter the posterior part of the eye through the pupil. The lens is normally crystal clear. When we look at a normal eye the pupil appears black because we see right through the lens into the dark posterior vitreous cavity behind it. When a cataract has matured, or *ripened,* the pupil becomes white (figure 6–1). Just as we cannot see through the pupil, the patient cannot see out of it. Only light and dark are discernible at this stage.

Progression of Cataract

The fully developed cataract is preceded by a progressive clouding of the lens, which occurs over a variable period of time. It usually takes years to go from the earliest detectable clouding seen with the eye doctor's instruments to a stage at which vision is significantly impaired, though the progression can occur over a period of months. Severe eye trauma or an illness that greatly alters the body chemistry can cause the rare cataracts that develop over hours or days. Trauma or ocular disease can cause cataract in the affected eye only. Cataracts due to aging or systemic disease usually develop in both eyes, though occasionally, for no known reason, the rate of development in each eye may differ markedly.

Causes of Cataract

There are many other causes of cataract besides old age, though the overwhelming majority are associated only with aging. As described in chapter 3, the lens is a very delicate almond-shaped structure composed of concentric rings of protein fibers, looking much like an onion. As we age, more rings are produced around the outside but within a thin transparent capsule. The lens is thus an active, living structure that requires nutrition and produces wastes. Since it must remain clear to transmit light (just like the

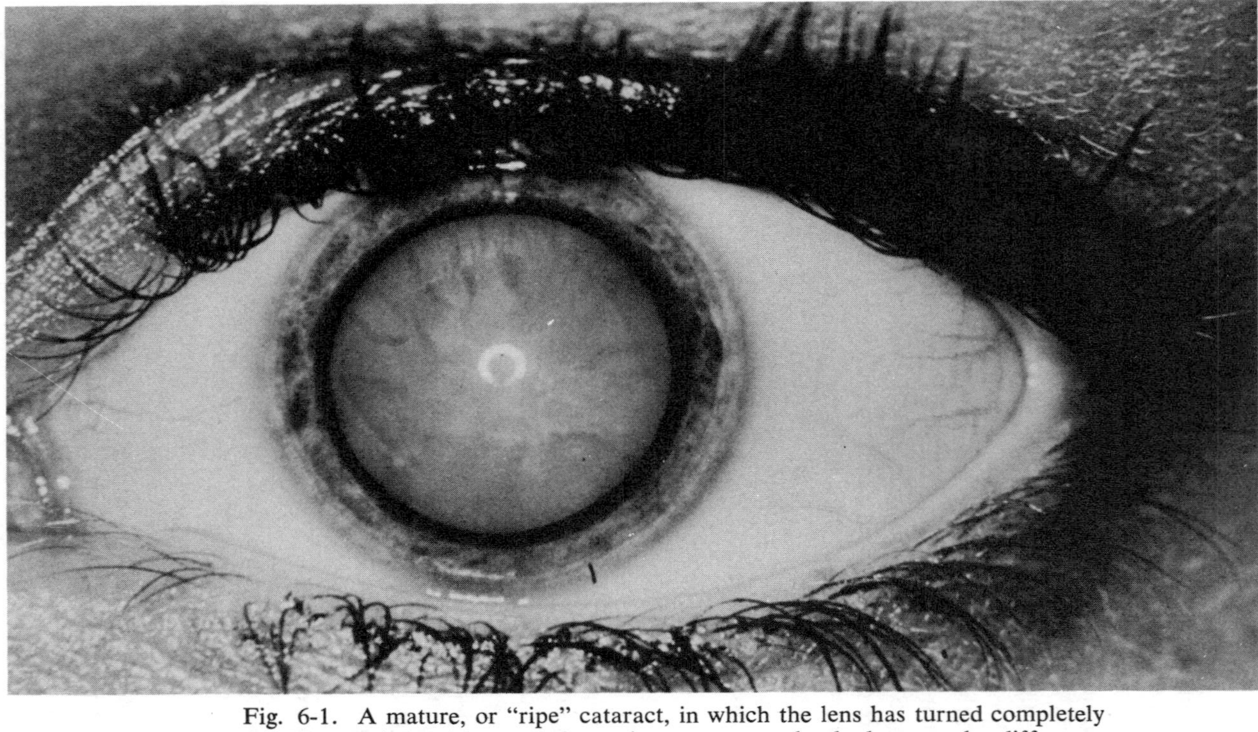

Fig. 6-1. A mature, or "ripe" cataract, in which the lens has turned completely white. The pupil is white, and the patient can see only shadows or the difference between light and dark through this dense cataract.

cornea), it cannot have opaque blood vessels passing through it. It depends on the aqueous humor, the surrounding clear fluid which circulates in the front of the eye, for delivery of nutrients and removal of waste products. Alterations in the normal chemical environment can alter the lens, which may result in denaturation of the clear protein strands and the accumulation of excess water in the layers of the lens. Such alterations are seen as clouding, or cataract formation. The chemical environment of the lens is known to change with underactivity of the parathyroid glands (which regulate calcium) and with a rare genetic disorder observed in infancy in which galactose sugar cannot be metabolized properly by the body due to a deficient enzyme. These very rare conditions can cause cataracts, as can a number of toxins, drugs, and other diseases that affect the whole body. Ophthalmologists always consider these possible contributing factors or otherwise unsuspected will direct dosages of radiation and electrical shock of sufficient severity. But, as noted, the vast majority of cataracts are due to poorly understood factors associated with aging. In cases of severe but nonpenetrating blow to the eye will cause a cataract, as will direct dosages of radiation and electrical shock of sufficient severity. But, as noted, the vast majority of cataracts are due to poorly understood factors associated with aging. In cases of definite calcium or galactose imbalance, early correction of the metabolic problem may arrest and even reverse the cataract formation. But these cases are literally one in a million.

We have yet to find any consistent abnormal substances or deficiencies of normal substances that could be used as a basis for rational medical therapy. Thus no known vitamins, diets, or pills represent an effective treatment for cataracts.

Cataracts are found in all age groups, though they are rare before the sixth decade of life. Some infants are born with cataracts due to a genetic tendency, or as part of a generalized disease process, or from maternal infections during pregnancy. *Rubella,* the German measles virus, is a prenatal infection that can cause cataracts. Cataracts can develop at any stage in life from these same basic processes: a genetic tendency; an ocular disease or trauma; or a generalized disease, toxic factor, or drug that also affects the lens.

Symptoms of Cataract

The main symptom of a cataract is diminished vision. Just as a camera will take a poor, cloudy picture when its lens is dirty or defective, the retina, too, will perceive poorly focused and distorted light rays when the eye's lens is cataractous. As the cataract progresses, vision becomes worse, until, once the cataract is mature and the pupil appears white, all one can see is the difference between light and dark. Earlier on, other symptoms associated with aging may develop. As the lens ages, it becomes more powerful optically, but it also becomes harder and loses its ability to change shape in response to the need to focus at close range. Thus a normal sighted individual approaching his mid-forties needs reading glasses to focus up close at fine detail. As the lens hardens with age, its natural focusing power diminishes further and stronger reading glasses are needed to compensate. This normal accompaniment of middle age, called *presbyopia,* will receive more attention in chapter 12.

The earliest signs of a cataract may be an abnormal increase in the refractive power of the nucleus, or hard core, of the lens. The individual thus becomes nearsighted and may suddenly be able to read again without glasses. This phenomenon is referred to as "second sight," but unfortunately it indicates only improved close vision; distance vision will now be somewhat reduced and will require correction with glasses. In the most common instance, then, a person with normal vision will become nearsighted and lose distance vision. For the same reason, an already nearsighted person will become more so and a farsighted person will become less so.

It must be emphasized again that such symptoms are only representative and do not apply to all cataracts. Most importantly, they do not herald an inevitable or rapid loss of vision. The development of a cataract is an extremely variable phenomenon, taking months in rare instances, more usually several years, and occasionally more than ten years. Clearly, not all people with evidence of cataract need treatment; most are not even aware of its presence. With the few extremely rare exceptions to be mentioned later, a cataract is only as significant as it is symptomatic. That is to say, if one can see well enough, the cataract does not require operation.

The clouding of the normally transparent lens does not appear uniform and diffuse until the cataract is ripe. Initially, the nucleus of the lens hardens and acquires a yellowish or brownish hue, becoming progressively more dense and more difficult to see through. Localized cloudy areas may appear in the surrounding layers of the lens, called the *cortex*. Though they may be quite extensive, these focal opacities will cause no problem with vision so long as they are peripheral in the lens. A central opacity of the lens, however, called a *posterior subcapsular opacity,* can be quite disabling because the opacity, though small, is directly in the visual axis (figure 6–2). The pupil normally constricts for near focusing as part of a three-part reflex called accommodation. (The other two parts are the focusing of the lens in the eye, brought about by contraction of the ciliary muscle inside the eye, and the convergence of the two eyes onto the near target.) With a small pupil no light can get around the central opacity, so vision is markedly worse for near than for distance. For the same reason, vision may be much better at night or in darkly illuminated rooms, when the pupil dilates, than it is in bright sunlight, when the pupil constricts. The more common, diffuse cataracts may produce just the opposite effect. More intense light penetrates the cataract better, so these people see much better with highly contrasted objects. They do poorly at night or in dark surroundings, even though their pupils then dilate to let in more of the available light.

Dilating the pupils with eyedrops has long been a treatment for cataract. Regardless of the type of lens opacity, more light will usually improve the vision somewhat. However, glare may be quite disabling. Just as flecks of dust and grease and irregularities in a pane of glass will cause light to disperse in a dazzling pattern, so the opacities in an early cataract may cause glare and sensitivity to light. The first symptom of a cataract, then, may be reduced tolerance for light. At night, lights may appear to be surrounded by halos, and one may be unable to drive because of the glare from oncoming lights. Such problems are often minimal. They are

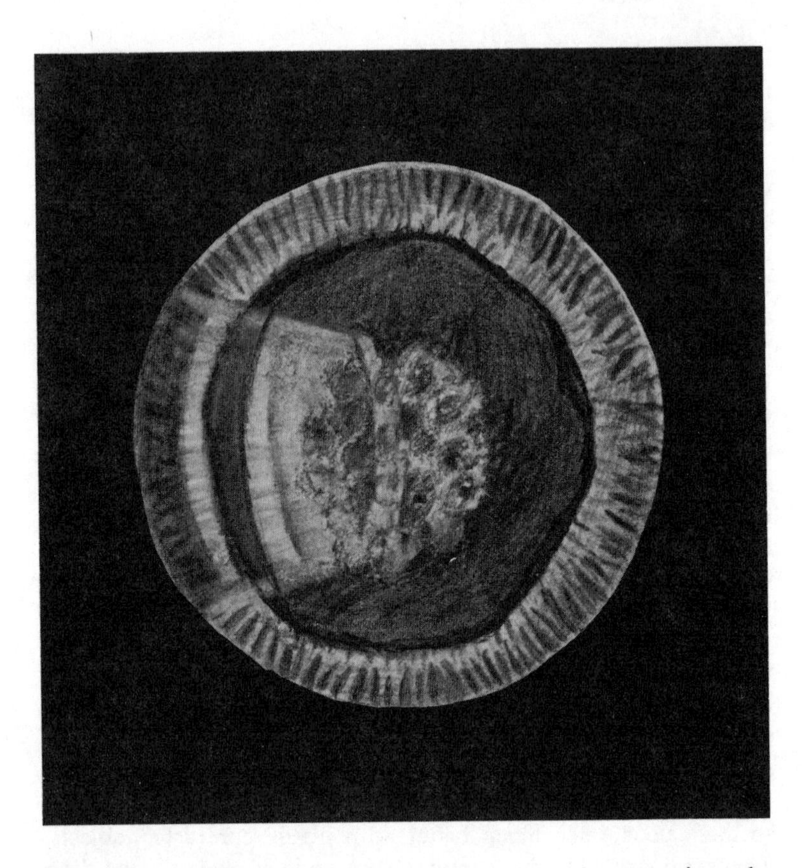

Fig. 6-2. Representation of a slit lamp view of a posterior subcapsular cataract in which cloudiness occurs within the posterior capsule of the lens, often centrally.

evident at night in people with peripheral opacities in the lens, because only in darkness will the pupil dilate widely enough to allow light to hit and be dispersed by these peripheral spots.

Artists and very discriminating individuals may notice as a first or early symptom a gradual shift in the color vision in the affected eye. The yellowish pigment of the cataract absorbs blue-violet light and robs these colors of their intensity.

These symptoms, then, represent the visual problems of a cataract and in almost all instances the only medical significance.

While light sensitivity and glare might be quite disabling, the real reason for removing a cataract is that the patient can no longer see well enough, and he cannot do what is important to him. Only very rarely will a cataract cause glaucoma or inflammation in the eye, requiring that the cataract be removed even though vision may still be at a tolerable level. A situation like this is more likely to result from a mature or traumatic cataract, but even then is uncommon.

Cataract Operations

While there are numerous variations on the techniques of cataract operations, the common factor in all is removal of the lens. Unfortunately, we have no way to strip the lens of its opacities and leave a normal, functional clear lens behind. As indicated, the opacity, or "skim," is the denatured and irregular lens substance itself.

Couching is the age-old technique of removing the opaque lens from the pupillary axis by using needles to push it down into the back of the eye. In this position, however, the lens may act as a toxic foreign body and cause severe inflammation, glaucoma, retinal detachment, and other problems. The complication rate of such a maneuver is so high as to make it totally unacceptable.

Intracapsular and Extracapsular

An incision must then be made in the eye so that the lens, or at least most of it, can be removed. Unless unusual conditions coexist, the safest and most convenient place to make the opening into the eye is along the superior border of the cornea, an area called the *limbus* (see figure 6–3). If the lens is removed totally intact within its capsule, the operation is called an *intracapsular cataract extraction*. When only a portion of the lens is removed, the procedure is termed an *extracapsular extraction*. The capsule is incised and the nucleus and as much cortical material as possible are removed.

While intracapsular extraction is still the prevailing technique, recent innovations in extracapsular extraction have excited new interest in this procedure. The essential difference from the patient's standpoint is that an intracapsular extraction removes the entire lens, leaving a clear visual axis with no chance of recurrent

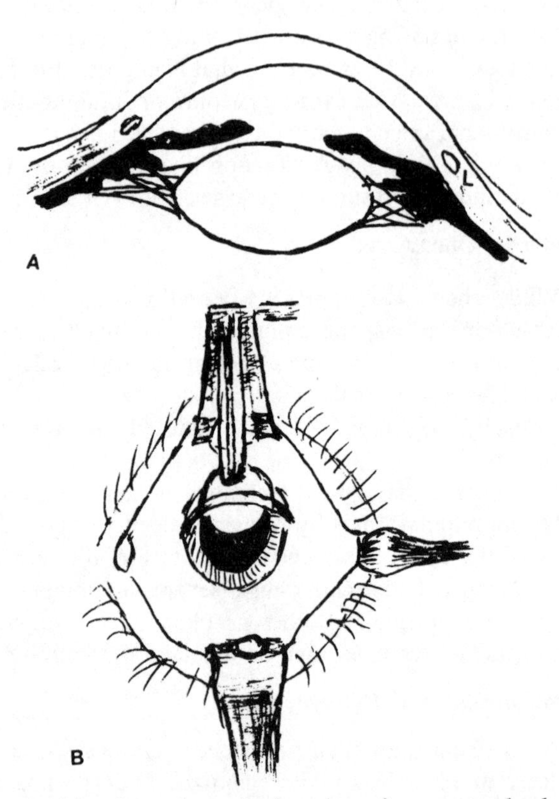

Fig. 6-3. An incision is made at the junction of cornea and sclera, an area called the limbus (L). In an intracapsular cataract extraction the lens is removed intact through the incision.

opacification. Extracapsular extraction leaves behind the posterior capsule and inevitably some cortical material. These may cloud up in the visual axis, requiring a second procedure to clear the area and restore vision. Usually all that is required is a simple momentary slashing of the capsule, called a *discission*. The tendency to develop progressive clouding of the posterior capsule diminishes with age. It is so prevalent in children and young adults that many eye surgeons do a discission at the time of the original operation to avoid the necessity later on.

Why do an extracapsular extraction at all? It was utilized initially because doctors thought it would be safer to leave the posterior capsule intact in the days when sutures were not available to close the incision. The procedure prevents loss of the vitreous, the gelatinous material filling the back of the eye, and it allows a smaller incision than that needed to remove the intact cataract by the intracapsular technique. Once adequate suturing techniques were developed, intracapsular extraction became the preferred method because it invited fewer complications.

Extracapsular extraction, however, is felt to be preferable for children and young adults because the nucleus is still soft enough to be aspirated through a simple needle. Also, the posterior capsule remains attached to the vitreous gel behind it, and loss of this gel is too likely with the intracapsular technique. This attachment is usually absent by middle age. Vitreous loss at operation may lead to greater postoperative inflammation and a higher risk of retinal detachment. A larger incision can result in greater astigmatism, or irregularity, of the cornea. This would then have to be compensated for with glasses, which might be difficult to get used to. Also, the larger the incision, the weaker it is and the longer it takes to heal.

Thus, in spite of an impressive success rate in the high 90-percent range, improvements in technique were still sought. Ideally, if one could not dissolve a cataract or clear it with medicine alone, an operation could be developed to remove the cataract through as tiny an opening as possible. Astigmatism produced would be minimal, and the small wound would be quite strong. With little danger of rupture to the incision site, convalescence could be rapid. The technique of *phacoemulsification* was thus developed.

Phacoemulsification

A sophisticated and costly ultrasound machine called the phacoemulsifier allows the eye surgeon to insert special probes into the eye to emulsify the hard portion (the nucleus) of the typical cataract and suck out (aspirate) this emulsified material along with the softer portions of the cataract. The advantages of this procedure have been already mentioned, as a small 3- or 4-millimeter incision is all that is necessary to introduce these special probes. The disadvantages of the technique center mainly about the fact that it is an extracapsular removal; a repeat operation may thus be necessary to produce a clear visual axis. Eyes with already weakened corneas, which could be due to a fairly common degenerative condition, endothelial dystrophy (see chapter 4), are perhaps more prone to accelerated progressive corneal problems; and certain types of cataracts, especially those with extremely hard centers, require so much ultrasound energy and emulsification that the eye is probably more traumatized than it would be with the more conventional procedure. When phacoemulsification is indicated and works well, however, it is a pleasure to have patients rehabilitated rapidly with a small wound.

Microsurgical Techniques

Profound changes have taken place during the evolution of intracapsular cataract extraction as well. More eye surgeons have begun to use the operating microscope for cataract removal, and excellent new sutures and instruments have been developed for use with this *microsurgical technique*. Intracapsular surgery can now be performed with great precision, creating strong, meticulously sutured incisions. Thus, rapid and safe rehabilitation with minimal corneal astigmatism can be accomplished with intracapsular technique as well. The common factor contributing to the progress of the various approaches to cataract surgery has been the introduction of sutures and instruments that allow more meticulous technique. In fact, the results of both modern intracapsular cataract extraction and phacoemulsification enjoy comparably favorable results. For decades, ophthalmologists have been debating whether or not an operated eye is better off with or without a remaining posterior capsule (the back capsule of the lens), and the question is still unresolved. The rationale for leaving this part of the capsule in the eye was that it might lower the incidence of retinal detachment and control macular swelling that might occur after a cataract operation. To date, there is no hard, widely accepted evidence to support this view; and it is known that this capsule usually cannot be left intact anyway, since it clouds up and has to be incised to permit good vision.

Routine extracapsular extraction was thus unusual (except as performed with the phacoemulsification technique) until the recent development of artificial intraocular lenses. A whole new area of investigation and discussion has developed around the use of these artificial plastic lenses to replace the cataractous lens. They can be inserted into the eye with either intracapsular or extracapsular techniques. From the patient's standpoint, this discussion within the profession is largely irrelevant right now; although if extracapsular technique is utilized, after-cataracts requiring reoperation are a potential side effect. Intraocular lenses are the latest innovation for dealing with problems of adjusting to *aphakia,* the term used to describe an eye that has been operated upon and now has no lens of its own with which to focus.

The Patient's Operative Experience

Cataract removal can reasonably be termed minor surgery. The operation itself takes less than an hour. Preparation of the eye and administration of anesthesia bring the total time up to about an hour. Local anesthesia, numbing medicine injected quickly near the eye and eyelids, is commonly used. Prior to coming to the operating room, where these injections are administered, the patient receives oral or injectable medicines to induce drowsiness, decrease anxiety, and dull the momentary pain of the injections. These medications may be supplemented as necessary during the operation; many eye surgeons even have an anesthetist in attendance to monitor the patient's level of anxiety and general condition during the procedure. Nevertheless, once the initial injections have been administered there is no significant pain, and the risk in undergoing a cataract operation with local anesthesia is minimal.

Under general anesthesia the patient is unaware of these preparations. Here, too, premedications are administered to ease anxiety and produce drowsiness. The patient then is put to sleep in the operating room by an anesthetist who will remain in attendance until the patient awakens after the operation.

Postoperative convalescence will vary with the surgical technique and the doctor's advice, but the patient is usually up and around within a day. After the operation, the patient can request mild or moderate pain medication, although it is not likely to be needed after the second or third day. No longer, as in the days before suturing and precise incisions, is one bedridden for weeks with the head immobilized by sandbags. Probably the patient will be told to refrain only from real exertion. Bearing down to have a bowel movement places a stress on the recently operated-on eye, no matter what operative technique was used. Laxatives can be taken if necessary. As added protection against bumping or rubbing the eye while asleep, a firm metal or plastic shield can be placed over the eye. Such precautions are rarely necessary after several weeks. The patient must exercise good judgment, which means consulting the doctor whenever in doubt. Even an obvious question is preferable to unnecessary complications or anxiety.

A general physical evaluation always precedes a cataract operation so that all potential risk factors can be discovered and assessed. Even elderly patients with infirmities have had cataract operations with little or no risk to their general well-being. Nonetheless, as with any operation or procedure, complications occasionally arise. These can range from hemorrhage to infection, from secondary glaucoma to retinal detachment. The doctor will discuss these problems with the patient and indicate whether the preoperative evaluation suggests any tendency for particular complications. Overall, however, the success rate of the modern cataract operation is excellent and among the highest of all operations performed. Approximately 95 percent or more of patients undergoing the operation can expect restoration of useful vision without any significant complications.

Rehabilitation and Restored Vision

We must now discuss the postoperative course and what one can expect visually from the successfully operated eye. With the exception of wearing a protective shield during sleep and applying drops or ointments to the eye for several weeks after the procedure, little irritation or annoyance is to be expected. Exactly what visual result to anticipate is a far more complex issue. To a large extent, what one sees after the operation and how one interprets this result depends greatly on the expectations of the patient, the preoperative state of the cataract and level of vision, and the visual needs of the individual.

Recall that removing a cataract is really removing the cataractous lens of the eye. The eye is not at all the same as it was prior to development of this cataractous clouding of the lens. A significant element of the eye's focusing system, the only element that can vary or adjust the focus, has been removed, so an entirely new optical situation exists. Without some artificial lens to replace the human lens, vision will remain blurred, but those who have had advanced cataracts will notice an immediate improvement in their vision. Some individuals, especially those in underdeveloped countries, are far less visually demanding than others. Cataracts often advance to near-maturity before help is sought. After having been practically blind, these patients notice a dramatic improvement in their ability to see and to get around after cataract removal. Many will not bother returning for corrective lenses to improve acuity, as their visual requirements have been met.

Others will not be so satisfied. Often the cataract represents a significant occupational or reading problem long before it is so far developed that it actually impairs one's ability to recognize large objects or to get around alone. Unless a significant cataract is present in one eye only, it is rarely allowed to become so advanced. Thus, without final optical correction, vision in the newly operated eye is likely to be considerably worse than it was prior to removal of the immature lens. An exception to this rule is the case of a previously very nearsighted eye. Such an eye is optically nearsighted because the greater length of the eyeball or more convex shape of the cornea actually render it too powerful in an optical sense (see chapter 12). Removing the lens of the eye significantly reduces this excessive power, and patients will find that they see better without glasses than ever before and that they may need none for either distance or near vision. Probably they will wear thinner glasses than they have ever used. Such instances are rare, however. The vast majority of patients obtain visual correction after cataract operation either with spectacles or with contact lenses.

Cataract Glasses

The methods of visual correction that can be used after successful cataract extraction include spectacles, contact lenses, and intraocular lenses. We have already mentioned that cataract removal is removal of the lens of the eye. The otherwise normal or near normal eye needs a replacement for the optical power that this lens within the eye provided. The most commonly used correction are spectacle lenses. Of necessity, the lenses are usually quite thick and the glasses heavy. Plastics and new lens designs have reduced the thickness and weight somewhat. But of greater importance and concern to the patient initially is the totally new type of vision these glasses provide.

Because the strong lens within the eye has been replaced with a lens that sits outside the eye and approximately ½ inch in front of it, a whole new optical system is created. In a sense, it is like looking through the opposite end of a telescope. Rearranging the position of this lens from behind the cornea (and within the eye) to the front of it creates significant magnification. Thus, images through the new cataract glass appear to be about 30 percent larger than they did before. The new, bigger world as seen through these glasses may pose an initial adjustment problem, to which one almost always easily adapts. Of greater consequence initially is so-called *spherical aberration,* a term referring to the distortion that thick lenses produce when off center. Patients frequently report that a doorway in the distance appears to bow in at the center. As they approach it, the sides seem to separate and straighten. Motion can add to this misperception. A straight row of utility poles may seem like putty rods in an earthquake, or a large building with columns may appear in a state of active collapse. These effects can be negated by keeping the eyes basically motionless and gazing only through the optical centers of the lenses. To look from side to side, one must now move the head rather than the eyes. Newer types of so-called aspheric lenses compensate for some of the distortions and the power changes needed to focus as the eye moves off the center of the lens, but these aberrations as yet have only been lessened, not eliminated. One can understand, then, that new reflexes must develop and new interpretations of vision by the brain must be "programmed in."

While this process goes on, simple manual tasks may be quite difficult and awkward. Depth perception is impaired and the location of objects may be misjudged. One must be prepared for minor accidents, such as spilling a cup of coffee, or reaching for an object in what is actually the wrong direction. With practice, though, vision will become near perfect again. Having overcome this initial clumsiness, motivated people will find themselves once again engaging in normal activities in a surprisingly short time. As with any adaptation to a new situation, though, there is enormous individual variation depending on personality, acceptance of the initial problem, desire to overcome subsequent problems, and most certainly the actual visual needs of the person after the operation. Some people will be thrilled from the beginning and never experience a problem with the glasses. At the other extreme is a very small group of people (fortunately) who never adapt. They may use the glasses for reading only. In between is the vast majority who do notice the problems, experience some initial difficulty, but within several weeks to months fully adapt. Of course, the greater the visual needs and discrete eye-hand coordination activities required for one's return to usual activities, the more adjustment and practice will be necessary.

Once the initial adjustment to the glasses has been made, two problems will unfortunately remain as the price of vision restored with spectacles. One involves the effect upon one's visual field of a large magnifying lens out in front of the eye. The visual field is the term applied to the entire area we see in space. Normally we see all that is directly in front of us. In addition, we see and are aware of things off to the side and up and down. This peripheral, or side, vision is of great use in orienting us and helping us to avoid collisions with things that are in our way but are not directly in our line of vision. It also serves as a warning of approaching objects or persons. Just as a telescope magnifies what we look at directly but restricts the visual field, or total amount we can encompass in this magnified view, to a lesser extent so do cataract glasses. They are somewhat more confusing, however, in that one sees the magnified view straight ahead but still has some far peripheral vision from around the edges of the cataract lens. Thus a peripheral field is seen around a magnified, superimposed central field. Because the thick cataract lens magnifies, this central field overlaps the outer peripheral field; they do not coincide exactly.

The result is a ring of visual field in which one does not see. At close range, as in reading, it is not noticeable. Also, at distances greater than about eighteen feet it is rarely of significance, since the central field is wide enough. At intermediate ranges of three to ten feet, however, the size and location of this nonseeing ring of vision can be most annoying. It creates a phenomenon referred to as "jack-in-the-box." While looking at an object or person in a room, someone may suddenly pop into view quite nearby, seemingly out of nowhere. More importantly, while crossing a street one must watch carefully for traffic before, as cars approaching from a distance may suddenly disappear. These cars or others may then reappear suddenly and frighteningly close. Moving the head to keep track of the whole field is necessary. Although new kinds of aspheric lenses have also reduced the extent of this problem, it has not been eliminated.

The other constant problem one confronts with thick cataract glasses is keeping them properly adjusted and aligned. The power of such thick lenses varies significantly with their distance from the eye. Thus, even a one-millimeter slide down the nose, or a push of the glasses closer to the face, can alter the clarity of the focus considerably. Also, a vertical or horizontal malalignment of the optical center of the lens so that it does not coincide exactly with the pupil can introduce not only distortion but apparent displacement of objects in space due to a prism effect. Moved off the visual axis, spherical lenses become pyramidal lenses (prisms) which angulate light rays to produce an apparent displacement of images. If one lens is off-center, double vision can result. If one has bifocals rather than separate reading glasses, the bifocal segments must be aligned exactly for the same reasons. The ophthalmologist will take critical measurements for cataract glasses and let the optician who fits the frame know exactly for what distance from the eye the prescription has been determined. The optician can then calculate any necessary changes depending on the fit of the frame that is chosen. Of course, it must be carefully checked afterwards, but adjustments will become a way of life if critical vision is desired. The spectacles may fall from the night table, or the earpiece may loosen slightly, so a spare pair of glasses can be most helpful until adjustments or repairs are made.

A Significant Cataract in One Eye Only

From this discussion of the new visual world created by spectacle correction after a cataract extraction one can perhaps understand the problems that occur with spectacle correction after cataract operation in one eye only when the other eye is normal. For patients with normal or near-normal vision in one eye, spectacle correction will probably be unacceptable. Correcting one eye with a thick cataract lens will, of course, lead to an odd-looking and unbalanced pair of glasses, at best. But more importantly, the vision in the operated eye will be magnified about 30 percent more than that of the opposite eye. One will thus experience double vision, a smaller inset of the same view superimposed on a larger one from the operated eye. Very few patients can tolerate this great disparity by having the brain adapt and fuse the two perceptions into a coherent picture. A contact lens or intraocular lens will eliminate this problem, but unless one is motivated to pursue these alternatives, it is quite unnecessary and of no visual help whatever to remove even a moderately advanced unilateral cataract. For elderly individuals, or for those who do not need critical three-dimensional vision, it simply is not worth the risk (though minimal) or inconvenience, unless the cataract is an extremely rare one that creates glaucoma or inflammation in the eye. And as long as the ophthalmologist examines the eye regularly to be sure no unusual problems develop, the cataract will do no damage to the eye. The potential for restored vision will remain the same whether the cataract is removed early or late in its development. A "ripe," or mature, cataract is more likely to cause complications if it remains in the eye, but even these can usually be left untreated indefinitely without problems.

One additional problem, however, can develop when a cataract has become very advanced in one eye only. Since the vision in the cataractous eye is poor and indistinct, the eye may drift out of alignment with the other eye. Double vision will not result if the eyes become crossed or out of line due to the very poor vision in one, so there is no stimulus to keep them aligned. If such an advanced unilateral (one-sided) cataract (or any other problem severely affecting the sight of one eye) occurs before age five or six, the natural tendency of the poor eye is to cross inward; later in life the eyes tend to drift outward. Presuming one has such an advanced cataract in one eye that the eye can drift out of line without causing double vision, one must be prepared for the double vision that will result when the cataract is removed and vision restored. When eye doctors notice a definite tendency for such an ocular deviation, they may then recommend cataract surgery that might otherwise have been postponed. If the deviation is small, simply restoring the sight with consequent double vision for a day or so will be enough to stimulate the eyes back into alignment. The larger the angle of the deviation and the longer it is present, however, the more likely it is that measures such as prisms (special lenses that compensate for the angular deviation of the eyes) added to a spectacle correction, or additional surgery on the eye muscles will be necessary (see chapter 7).

If one is motivated by occupational needs, level of activity, or a developing ocular deviation, how can one correct the vision in the operated eye after unilateral surgery? The only satisfactory answers for the vast majority of patients are the contact lens or the intraocular lens. We will now discuss these two modes of correction of aphakia.

Contact Lenses after Cataract Extraction

Contact lens correction of aphakia is far superior optically to spectacles. Because this lens rests upon the eye and floats on a tear film layer over the cornea, it eliminates the problems of peripheral distortion, loss of peripheral visual field, and the ringlike area of nonseeing that plague the spectacle user. Also, the magnification from contact lens correction is not as great as that obtained with spectacles. A contact lens in this situation will magnify only about 8 percent, an amount that most people find tolerable and unnoticeable. Thus, when a cataract is removed from one eye and vision in the other remains near normal, one can wear a contact lens on the operated eye without experiencing the double vision we described in the case of spectacle correction. The contact lens can restore full, near normal binocular function. And, of course, the prospect of better overall vision makes contact lenses desirable for many people who have had both eyes operated. For a well-motivated individual, contact lenses represent excellent visual rehabilitation after cataract surgery.

Unfortunately, not all people can wear contact lenses. Psychological fears and the discomfort of having to manipulate something in the eye are bothersome to some people. Stroke victims and people with arthritis or a tremor may be physically incapable of handling the lenses themselves and unable to rely on others for help with insertion and removal of the lens. Still others have certain medical conditions of the eyelids, such as chronic inflammations, allergies, or problems with the cornea or tears, which make wearing a contact lens dangerous or impossible. New and improved lenses are being developed constantly, and more and more people are able to wear them satisfactorily. Most awaited, however, is a contact lens that can be worn twenty-four hours a day. The eye doctor can insert the lens, after which it will need only to be infrequently checked or cleaned. There are some lenses available now that can be used in this way by some individuals, but unfortunately not by enough. Most people still have to remove the lens after no more than twelve to eighteen hours wear. Removal allows the cornea six or eight hours rest for unimpaired oxygen uptake and for exchange of nutrients and metabolic waste products. (See chapter 14 for a more thorough discussion of contact lenses and the different types available. The foregoing general comments apply to the hard, soft, and semisoft lenses presently available.)

The Intraocular Lens

Because of the significant number of individuals who cannot wear contact lenses for one reason or another after cataract surgery, and because of the optical problems that follow spectacle correction, the idea of replacing the cataractous lens with a clear, equally powerful artificial lens inside the eye has long been attractive. An artificial intraocular lens would immediately restore near-normal vision, as if a cataract had never been present. Of course, the lens could not vary its focus, but this problem is easily compensated for with bifocal or trifocal glasses. Naturally, most cataract patients are elderly and already accustomed to such glasses, since their own natural lenses have long since lost the ability to vary focus significantly (see chapter 12).

During World War II, a British eye surgeon named Ridley observed that fragments of a certain type of plastic, used in the canopy of fighter planes, caused little inflammation when they penetrated a pilot's eye. So began the use of a material that proved so inert and lightweight that it has yet to be significantly improved upon for intraocular use. The initial intraocular lenses, however, were not as well accepted by the eye as tiny fragments of the same material. Numerous eyes were lost in the effort because the problems of design and stabilization of the lens proved so formidable, and the idea was almost abandoned. But a Dutchman named Binkhorst renewed the efforts, redesigned lenses and was in large part the impetus behind the new wave of interest and enthusiasm for these lenses that has grown in recent years. Needless to say, however, such lenses are still not perfect. Some operative techniques adapted to use with intraocular lenses are still controversial. And the risk involved in having an intraocular lens implanted at the time of surgery must be considered greater than that in a routine operation without implantation. Figure 6–4 shows some of the many lens designs currently in use. They are positioned in the pupillary area by various types of struts, pins, or sutures designed to maintain fixation either to the iris, or to the posterior capsule and lens remnants if an extracapsular extraction is done. Some are stabilized anterior to the iris in the angle the iris forms with the cornea; while some newer models are implanted behind the iris.

One of the complications of an implantation procedure is potential dislocation of the lens. Sometimes the dislocation is minor and can be corrected by positioning of the head or by changing the size of the pupil with eyedrops. In other instances, it can be disastrous. If the lens falls back into the eye and creates inflammation or scarring, retinal detachment may result. If it dislocates too far forward and touches the inside of the cornea, it can destroy many of the cornea's critical cells. The result can be a swollen, cloudy, or opaque cornea (see chapter 4). Chronic inflammation and glaucoma may be triggered by an intraocular lens. Very rarely,

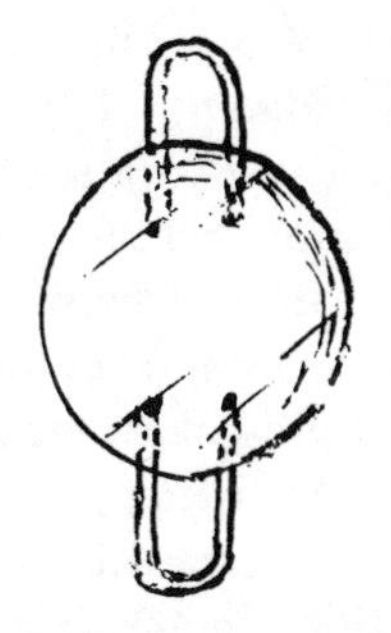

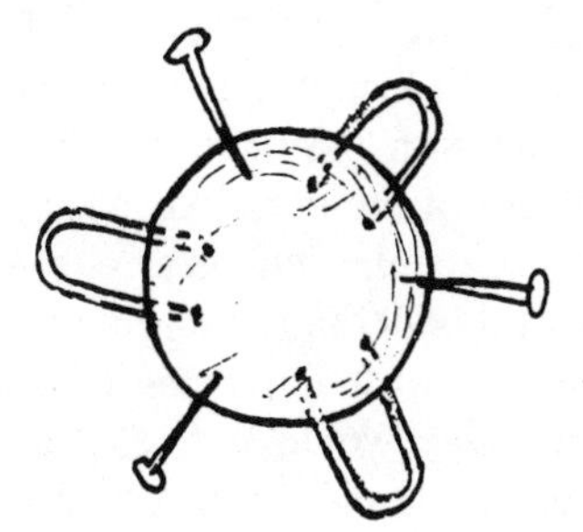

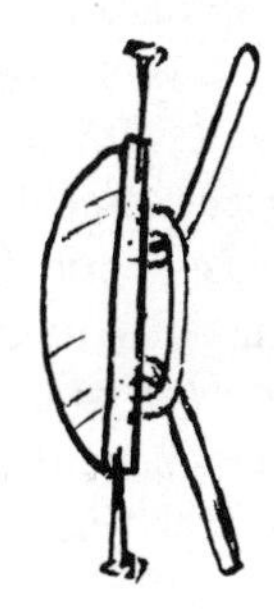

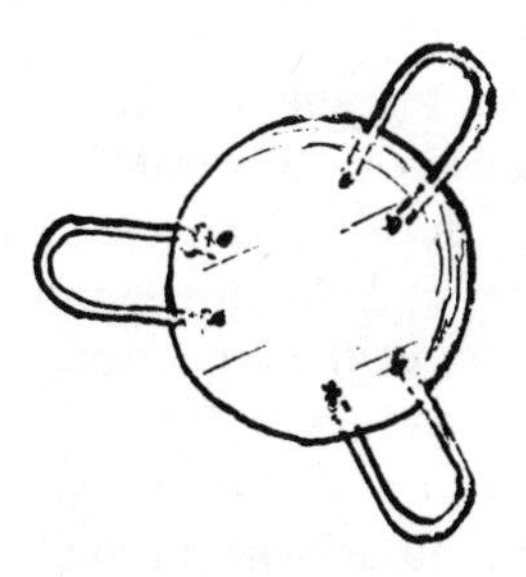

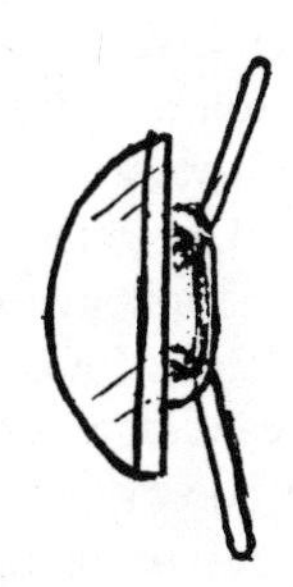

Fig. 6-4. Some designs typical of intraocular lenses presently employed.

infection can be introduced directly with the lens if contamination has occurred. In an extracapsular cataract extraction technique (an increasingly popular way of providing presumably safer fixation of the lens) secondary clouding of the posterior capsule or new membranes and attendant scar formation may develop across the pupil. An additional operation will then be required. Its magnitude and significance depends on the density and thickness of the membranes to be opened or cut free from the pupillary area.

Any operative procedure represents trauma, and the longer the procedure the more intense the trauma. Intraocular lens implantation probably lengthens the cataract operation and hence kills more cells from that critical inner lining of the cornea than does routine cataract extraction. The long-term implications of this finding are not yet clear.

In spite of all these problems, the proponents of intraocular lens implantation claim that complications are not significantly more frequent with lens implantation, particularly when one considers the great optical advantages of the intraocular lens. Many ophthalmologists agree that an intraocular lens is a reasonable alternative for an older individual who is definitely unsuited to spectacle correction (as would be the case with an advanced cataract in one eye only) and who most probably would not tolerate a contact lens. Most feel that these lenses should be reserved at this time for older individuals, since well-controlled, statistically significant studies have not yet been conducted on long-term (twenty years or more) effects of these lenses. But there are compelling reasons in some individual cases to make an exception, to take a greater risk.

Absolute scientific criteria for the use of intraocular lenses are still wanting. In the absence of statistically significant studies, isolated surgical triumphs or disasters can give rise to very adamant arguments. Certainly reputable, conscientious ophthalmologists can be found on both sides of the arguments over who should be a candidate for such lenses and under what circumstances. The more conservative eye surgeons feel that the advantages do not compensate for the increased initial risks, the unknown long-term risks, and the probability that a widely accepted continuous-wear contact lens will be developed shortly. Proponents say that the initial risks are minimally increased and that the long-term risks are very remote. They also point out that an extended-wear contact lens acceptable to most people (which would clearly eliminate the advantage of an intraocular lens in the majority of cases) has been expected for decades and has only recently been made widely available but not yet proven universally acceptable.

These extended-wear contact lenses are not tolerated by everyone. They may cause serious infection in some. They have to be removed at varying intervals for cleaning; some patients may be able to wear the lens for months, others for only a few days, before mucous coating and debris or irritation require removal for cleaning and/or a respite. Wearing such contacts mandates a committment to indefinite, close follow-up by one's ophthalmologist.

Thus, there are eye surgeons who implant lenses in both eyes of patients with cataracts, despite the age; others consider implanting a lens only in quite elderly individuals, and then in one eye only. Of course, a spectrum exists in between. It should be mentioned that many experienced intraocular lens surgeons will not implant a lens in an eye that has had an intracapsular cataract extraction at a prior time. They feel it is dangerous and the complications frequent. At this time, for most people who have already had a cataract operation and are still very dissatisfied with contact lenses or spectacles, artificial lens implantation in the operated eye must be considered a significant risk.

How, then, should one decide? Clearly, if one decides at this time against intraocular lens implantation and then has a cataract extraction without it, one must contend with available contact lenses or spectacles. But once a lens has been implanted, the wearer must be equally resigned to a long period of uncertainty and several additional short-term worries. A nationwide study underway in the United States presently should better quantitate the actual risks involved in lens implantation.

How to Decide on a Course of Action

First, one must be absolutely certain a cataract operation is necessary at all. If patients themselves are not convinced that their own visual status is sufficiently disabling or annoying to warrant the operation, then the ophthalmologist must have compelling medical reasons to suggest it. Several have been brought out in this chapter.

Once a decision that something must be done has been reached and accepted, considerations of the type of procedure and anticipated type of visual correction afterward should be entertained. Conservative people who are unwilling to risk their eyesight or who are psychologically upset by the idea of a foreign, plastic device inside the eye should not consider an intraocular lens. Those who are thrilled or intrigued by the prospect of having a cataract removed and using only ordinary glasses rather than thick cataract glasses or contacts might consider the intraocular implant. Some surgeons have hedged their bets in patients with bilateral cataracts by putting an intraocular lens in one eye and fitting the opposite with a contact lens. The near normal vision in the eye with the intraocular lens helps the patient to manipulate the contact lens.

Visual needs must always be the first consideration. Most cataract patients are elderly and do not pilot airplanes, perform critical surgery, or play professional sports. They most often adapt remarkably well to cataract spectacles, despite the disadvantages. And elderly or relatively undemanding patients with uniocular cataract, if it is adequately advanced to warrant removal, will still appreciate better side and overall vision after operation even without any correction for good central acuity in that eye. Individual patients must be prepared to tell an eye doctor what their visual needs really are and how rapid their rehabilitation must be. They must honestly confess and discuss the risks they are willing to take and what their feelings and anxieties are regarding the use of a contact lens, the presence of something foreign in the eye, and the cosmetic and social consequences of thick glasses. What they can physically manage alone or with the assistance of others in another important consideration for those contemplating contact lenses.

Armed with this information, one must sit down with a trusted eye surgeon and discuss the options that are available. The patient cannot make the decision alone because not all possibilities are appropriate for all individuals. The relative risks will vary from case to case. After a thorough eye exam, the ophthalmologist may tell the patient that certain possibilities are out of the question, or relatively more risky given the particular condition of the eye. One can consult other ophthalmologists if dissatisfied with such conclusions. But in the end, one must consider the physical suitability of the eyes for these alternatives. If opinions differ, one can either rely on consensus or on a single trusted and respected ophthalmologist.

While confusion and despair may at first overwhelm the cataract patient, optimism should prevail. Never before has such an excellent prognosis for restoration of useful, near-normal vision been available. The confusion results from the progress, which now suggests possibilities for even better results.

7
Strabismus and Amblyopia

While crossed eyes and lazy eyes are extremely common, misconceptions about these problems abound and often lead people to delay treatment until it is too late. Increased awareness of the disorders will lead to earlier detection and an improved chance for a cure.

Strabismus is the technical term for crossed eyes, and *amblyopia* is the term for lazy eyes. Many people confuse the two. Vision in the amblyopic eye is poor and cannot be corrected with glasses or contact lenses. Amblyopia is one of the ways in which a child compensates for crossed eyes in order to avoid double vision. But a crossed eye is not necessarily an amblyopic eye, for it may see perfectly when the opposite eye is covered; it may be that vision in this crossed eye is ignored by the brain only while the opposite eye is open and focusing on something. When the opposite eye is covered, the formerly crossed eye picks up fixation and has normal vision. It is not visually deficient in itself; its sight is temporarily turned off by the brain while both eyes are open. An amblyopic eye will remain deficient even when the opposite eye is covered and spectacles or contact lenses are worn.

Strabismus, or malalignment of the eyes, occurs in approximately one percent of the overall population, although its incidence is higher among preschool and school-age children. Amblyopia is present in about 2 percent or more overall. In addition to its association with strabismus, amblyopia may develop in an eye with a markedly different refractive error (farsightedness, nearsightedness, astigmatism) from its mate, or in an injured or diseased eye denied normal vision for a critical period of time. True amblyopia will not develop after age six or seven. *Suppression,* whereby the brain adapts to malaligned eyes or asymmetrical vision by temporarily ignoring the extra image while both eyes are open, may develop later in life.

Although strabismus can arise from many different disorders, it is usually congenital (the exact cause being unknown) and is obvious at birth or within the first few years of life. When it develops later, it is said to be *acquired* and has quite different implications. Very young children with strabismus devise a variety of adaptations to avoid double vision, but older children and adults with previously normal binocular vision are quite disabled by the double vision that accompanies a sudden onset of malaligned eyes.

Malalignment of the eyes can be *comitant* or *noncomitant.* In comitant strabismus the amount (angle) of deviation between the two eyes remains constant regardless of the direction of gaze (figure 7–1), and eye motion is not limited in any direction. This type of deviation is the most common congenital type. Noncomitant strabismus implies that the angle of deviation between the two eyes varies with the direction of gaze; the eyes may be perfectly aligned in one direction but will then progressively diverge or converge as the gaze shifts (figure 7–2). This type of strabismus is usually acquired and is due to malfunction or scarring of an eye muscle or the nerve to that muscle. It deserves immediate attention and thorough evaluation.

Esotropia (Cross-Eye), Exotropia (Wall Eye), Hypertropia

If the malalignment of the eyes is in a convergent, or inward, direction, it is called an *esotropia.* If the eye turns outward, or diverges, the condition is called *exotropia* (figure 7–3). When a vertical deviation of the eye exists, the higher eye is said to have a *hypertropia*; less commonly, the lower eye is designated *hypotropic.*

Figure 7–4 shows that each eye actually has six extraocular

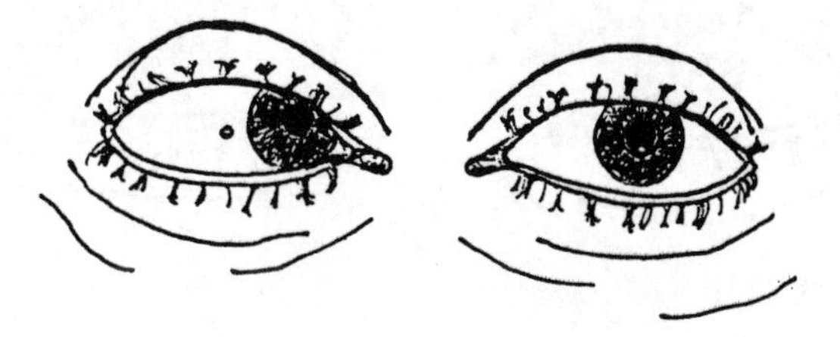

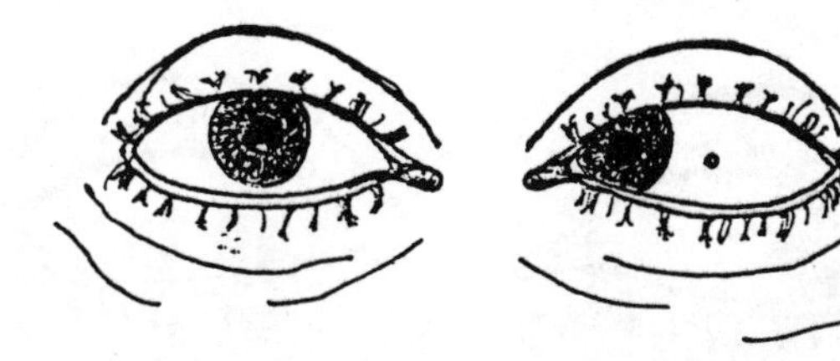

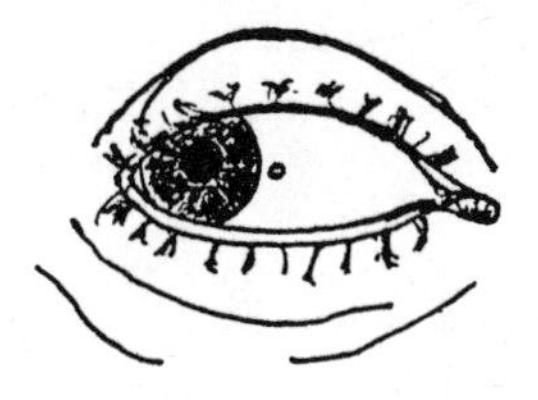

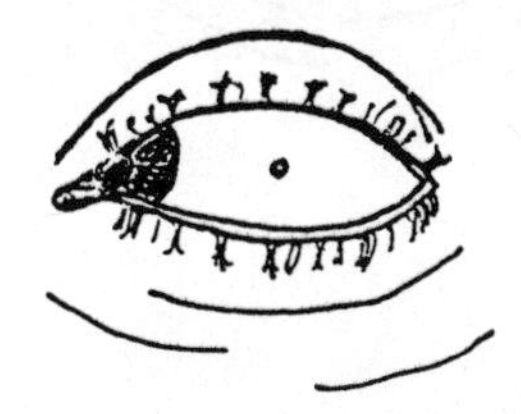

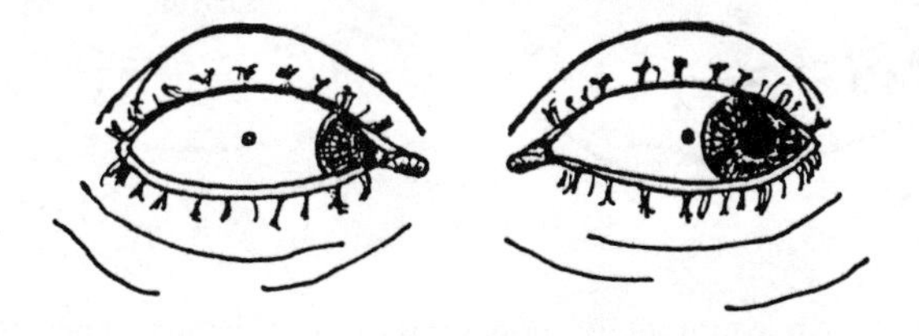

Fig. 7-1. An example of comitant esotropia. The eyes cross in a convergent direction and maintain the same angle of deviation in all fields of gaze.

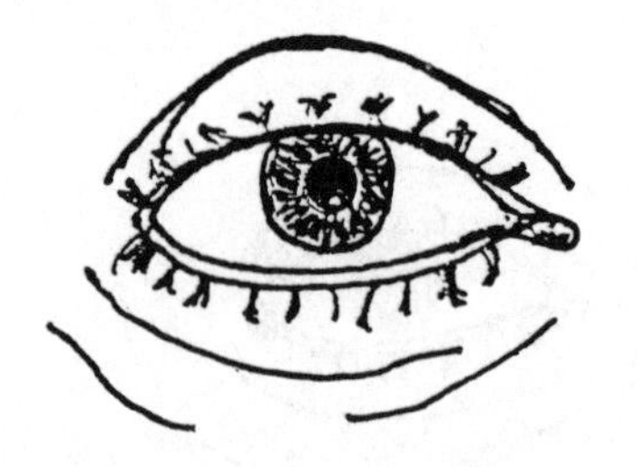
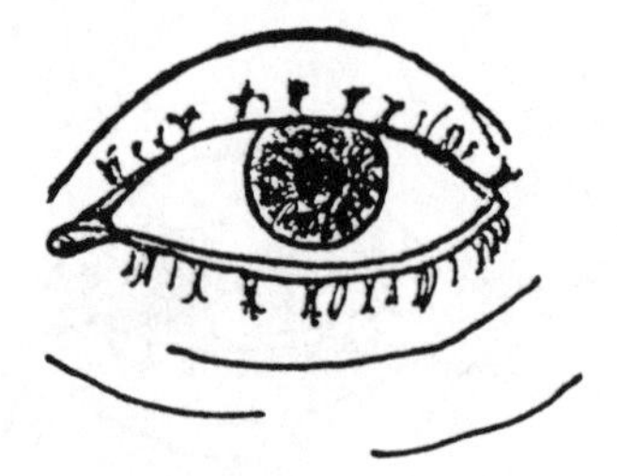

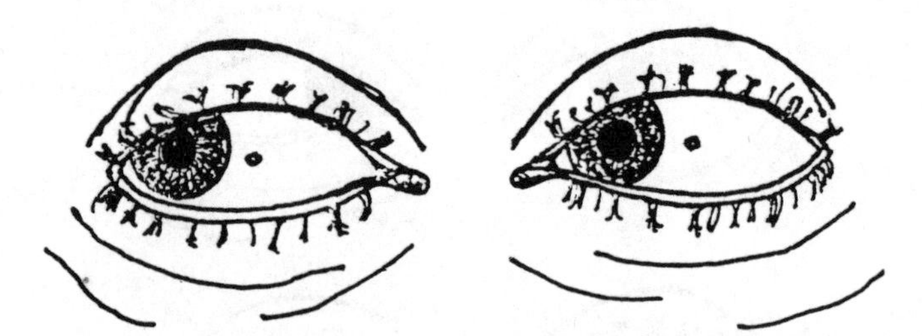

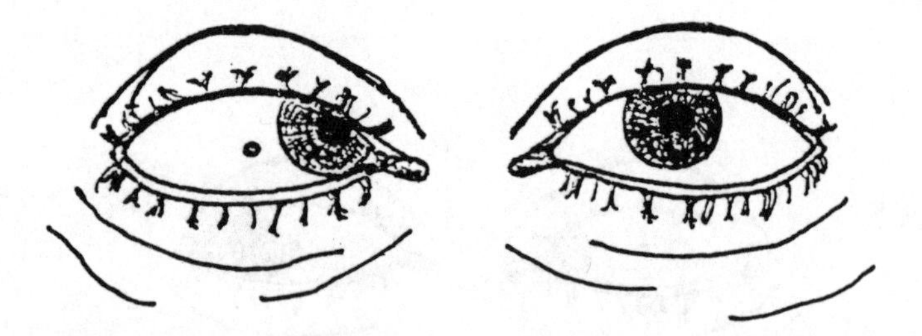

Fig. 7-2. An example of noncomitant strabismus. The eyes are aligned normally when looking straight ahead and to the right. In left gaze the left eye does not move normally, so the eyes cross.

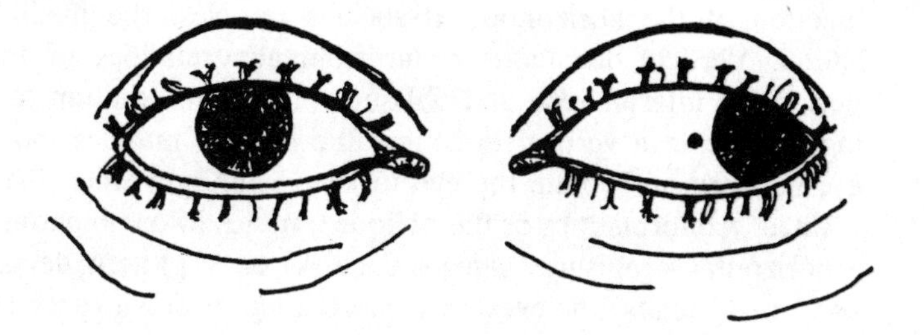

Fig. 7-3. In exotropia the eyes deviate in a divergent, or outward, direction.

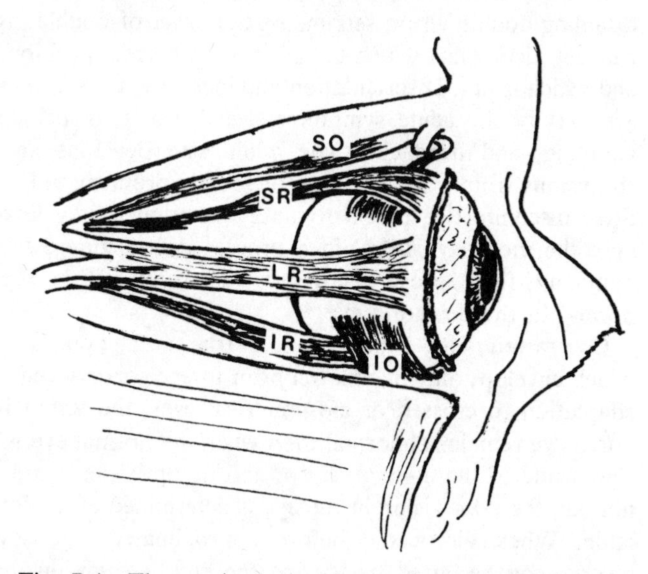

Fig. 7-4. The motion of each eye is controlled by six extraocular muscles: four rectus muscles and two oblique muscles. The drawing depicts the eye in the orbit as seen from the lateral (outer) aspect. The superior (SR), lateral (LR), and inferior recti (IR) and the superior (SO) and inferior (IO) oblique muscles are seen.

muscles—four rectus muscles and two oblique muscles. Horizontal deviations (simple exotropias and esotropias) can be related to the function of the horizontal extraocular muscles, the medial and lateral. Vertical deviations result from abnormalities of the superior and inferior recti and oblique muscles. In addition to moving the eye in a vertical direction, the oblique muscles move the eye outward and rotate the eye to keep it upright when the head is tilted. Abnormalities of the oblique muscles in combination with a horizontal strabismus can produce A- or V-pattern deviations, so named because the eyes cross more in up or down gazes (figure 7–5).

Sensory Adaptation

If one does not adapt in some way to malalignment of the eyes, disabling double vision sets in. At the onset of double vision, one can not distinguish which image is in its correct position in space and which is not. Disorientation and lack of eye-hand coordination are part of the acute symptoms, which can also include nausea, vomiting, and headache. The adult who develops an acquired strabismus from a neurological or muscle problem will experience these discomforting problems until the brain slowly learns to ignore the incorrect image, in a process called suppression. If the dominant, fixating eye is covered, vision will still be present and normal in the deviating eye.

Contrasted with suppression is true amblyopia, a condition which develops only in children prior to age six or seven. With this adaptation to crossed or asymmetrical eyes, the vision in the ignored eye remains abnormal even when the normal eye is covered. The inside of the amblyopic eye and its optic nerve are perfectly normal; the poor vision in the eye is determined at the level of the brain. When discovered before approximately age seven, amblyopia can be cured by forcing the child to use that eye. This treatment consists simply of patching the good eye until the vision returns to a normal level in the opposite eye. Patching may be necessary for as little as a week or two, or for many months. If the cause for the amblyopia is then corrected, the vision will almost always remain normal. Crossed eyes are one cause; another is markedly asymmetrical vision in two eyes. If one eye is very

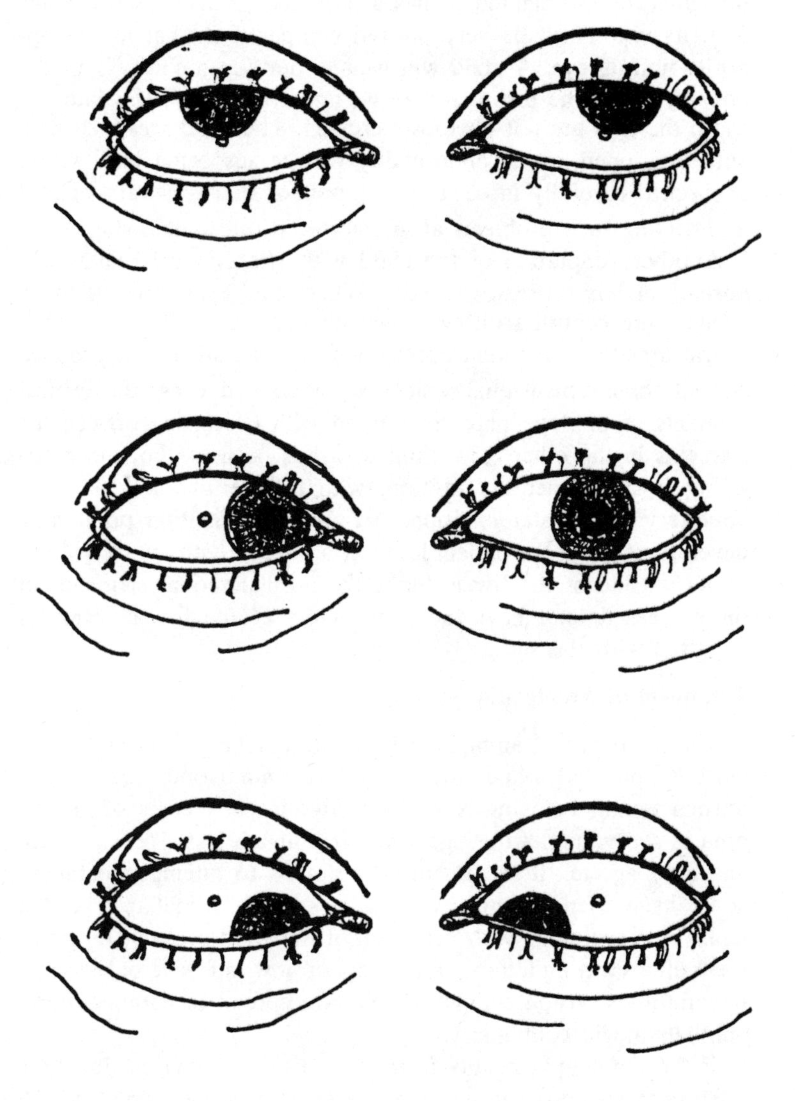

Fig. 7-5. A V pattern esotropia, so called because the eyes cross inward in primary position, more in a down gaze than in an up gaze.

farsighted or nearsighted or has a large astigmatism (see chapter 12), its vision will be very blurred compared to that in the opposite normal eye. A child will usually develop amblyopia in the poor eye, and the disorder may go undetected until too late because the eyes are not obviously crossed. The child sees well with both eyes open, so no abnormality will be suspected until vision is checked carefully in each eye. Preschool testing is very helpful in detecting such problems at an early and remediable stage.

Another adaptation of the child with strabismus is called *abnormal retinal correspondence.* When the eyes are perfectly aligned, the central fixating retinal elements as well as the peripheral areas of the retina correspond to each other and produce normal three-dimensional vision. With crossed eyes, the central elements of one eye may correspond with eccentric (off-center) elements in the other. The child will thus be using both eyes together in a rudimentary fashion, with full peripheral vision but imperfect central stereo vision. An abnormal fixation point may develop such that the patient looks at an object with an area of the retina outside of the fovea. Since the normal retinal elements in these areas do not have the good acuity of the fovea, vision is impaired with this eye.

Treatment of Amblyopia

The treatment of amblyopia is simple patching. The better eye must be patched while awake until the amblyopic eye regains normal vision. Patching is very effective in early cases of amblyopia in the young. After age seven it is much less likely to help, and after age ten it seems almost pointless to attempt treatment. Some have tried regimens of special sensory stimulation to the amblyopic eye along with occlusion of the good eye, but no treatment after early childhood, regardless of how intensive or extreme, has improved vision enough to gain widespread acceptance in the ophthalmologic community.

If the amblyopia results from very disparate images due to a large refractive error in one eye, the error must be corrected with spectacles or a contact lens. This type of amblyopia is called *anisometropic amblyopia.* If the amblyopia is a result of strabismus, once the vision is regained the eyes can be straightened by operation in the hope of real functional improvement. Abnormal retinal correspondence reverts to normal after the malalignment of the eyes is corrected.

Refractive Errors and Strabismus

Certain refractive errors can cause the eyes to cross, and it is imperative to rule out such conditions prior to any operative intervention. To understand how these problems must be handled, one must first understand the reflex called *accommodation*. When focusing on a near object, as in reading, three actions take place. First, the eye focuses on the object by contraction of the ciliary muscle (see chapter 3). This muscle causes the lens within the eye to achieve the proper focus. With this focusing process, the eyes automatically converge on the near object and the pupils become smaller.

In a farsighted eye, the image from a distant object comes to a focus behind the eyeball, just as would a near object in a normal eye. The farsighted eye thus focuses to bring this image onto the retina. But tied to the focusing is the convergence reflex—as when the eyes need to come together to look at a near object. The result is a tendency of the eyes to converge, or turn inward, caused by the farsighted eye focusing to see a near or distant object. This convergence reflex is strongest in young children, who have an enormous focusing capacity compared to adults. This capacity decreases with age (see chapter 3), as does the tendency to converge the eyes with the focusing effort. But in children, farsightedness (also called hyperopia) can thus cause crossed eyes. The only way to discover the refractive error of a child is to eliminate temporarily the voluntary and highly variable focusing of the ciliary muscle. The true state of the eye then can be determined by neutralizing with lenses the light reflected from the back of the eye. The light reflections are created with a hand-held instrument called the retinascope.

The Drop Test

The focusing ability of the ciliary muscle is eliminated with eyedrops called cycloplegics. The younger the patient, the greater his ability to focus, and the stronger the drops must be. Atropine is commonly used for this drop test in small children. In addition to uncovering any possible farsightedness, these drops dilate the pupils and can make the child very sensitive to light. A child should be restricted from bicycle riding or any potentially dangerous activity until the effects of the drops have worn off. In extreme but still normal instances, the pupils may remain dilated and vision blurred for a week or more after such a test. Usually the effects wear off within a couple of days if atropine is used, in only several hours if weaker drops are applied. Drops may be given in the ophthalmologist's office or, for very young children, at home. To maximize the effectiveness of the treatment, parents must administer the drops for several days prior to a return appointment. Very rarely will systemic absorption of the drops cause a child to get flushed, develop a fever, or act peculiarly, but should this occur, the drops should be stopped and the doctor notified immediately. Dosage is minimal, so the very rare atropine reaction will often last no more than a few hours.

Accommodative Esotropia

Any significant farsightedness should be corrected with glasses in a child with an esotropia. Often, the glasses alone will straighten the eyes. Surgery should never be performed for the amount of crossing attributable to farsightedness and accommodation. The child is likely to become less farsighted with age and will tend to converge less for a given amount of focusing. Thus operating for this degree of crossing will constitute an overcorrection later in life when the eyes may turn out to an exotropia!

Another facet of this same problem is found in children with an abnormal, exaggerated tendency to converge the eyes for a given amount of focusing close up. They may be significantly farsighted as well. But just the effort of focusing on a near object without additional focusing to compensate for hyperopia will cause the eyes to converge too much. Again, glasses will correct the problem. Bifocals are used to eliminate the extra focusing necessary for near vision, and the eyes remain straight. Surgery is not recommended for this condition either, as children invariably outgrow the problem. The strength of the bifocals can be reduced gradually until they are no longer necessary, usually by about age eight or ten. It is important, however, that the bifocals be appropriate for children. They must have full lower segments that cut across the level of the pupil. If given the small, inconspicuous bifocal segments preferred by some adults, children will avoid looking through them.

Eyedrops now are available to reduce or eliminate the tendency of the eyes to converge reflexly along with focusing. The drops are useful for very young or difficult children who will not tolerate glasses, but since the drops have minor side effects, most ophthalmologists prefer glasses. These eyedrops belong to the class of drugs known as *anticholinesterases.* A small amount of the drug may be absorbed into the bloodstream, even with infrequent applications, so individuals with asthma or heart conditions should avoid them. Also, a rare susceptible individual on this medication may have trouble recovering from a common muscle relaxant medicine used during general anesthesia, should any operation be performed and the doctors not advised about the eyedrops.

A general principle in the treatment and prevention of amblyopia is to provide the eye with as clear an image as possible. Thus a significant refractive error in any type of strabismus should be corrected with glasses, especially if the refractive error is different in the two eyes. Correcting all of the existing farsightedness in a child with an esotropia is always indicated, but in a child with an exotropia farsightedness will be undercorrected, while nearsightedness (myopia) will be fully corrected in order to stimulate the accommodation reflex and convergence of the eyes. Actually, overcorrecting with lenses for nearsightedness can help a child with a small exotropia. Lenses force the child to accommodate in order to see clearly, and in so doing he converges the eyes.

Prisms and Orthoptics

Prism lenses shift the apparent position of an object in space. When placed in spectacles, they force the eye to move in order to avoid double vision. Although such lenses have been used to treat strabismus, they are most commonly used for patients with impaired motion of the eye. Prisms, eye exercises, and sensory training are part of the discipline known as *orthoptics,* which is basically the nonsurgical treatment of strabismus. It is often helpful in the sensory preparation for surgery and in small-angle deviations, but it is not sufficient for most large-angle deviations. Large-angle in this context includes almost all eyes that are obviously maligned.

Surgery for Strabismus

Surgery is the prevalent and most successful means of realigning most eyes with strabismus. In the common situations, esotropia and exotropia, the horizontal rectus muscles are operated upon. By weakening and strengthening these muscles, the position of the eyes relative to each other can be changed. The common weakening procedure, called a *recession,* cuts a muscle free from its insertion on the eyeball and replaces it a measured amount further back along the globe, thus loosening it. The strengthening procedure is called a *resection.* Here, the muscle is cut free from the eye and a given amount excised at the end. It is then sewn back at its original site. The net effect is to tighten, or strengthen, the muscle. Thus, for an esotropia the surgeon recesses the medial, or inner, rectus muscle and resects the lateral, or outer, rectus muscle. Sometimes the medial rectus muscles of both eyes are recessed. The combination of muscles that is operated on depends partly on the angles of deviation while looking in the distance, at near, and in different directions. With vertical deviations and A and V patterns, similar surgery is performed on the oblique muscles or vertical recti.

For children, such surgery is performed under general anesthesia, and a one or two day hospital stay is usual. Even if the operation is done on an outpatient basis, the child must stay in a recovery room at least several hours until completely awake. Afterwards, the parents must understand how to care for and observe the child at home.

Although the eye may be somewhat red and require drops or ointment for up to several weeks after the operation, the recovery is not painful. The incisions to approach the muscles are made in the conjunctiva, the mucus membrane covering the eye. The scars are inconspicuous and are rarely complicated by cyst formation or irregularity (which can be corrected later if necessary). These conjunctival incisions heal within a couple of days, but the eye may remain slightly irritated from sutures that will not absorb completely for up to several weeks. It is wise to avoid swimming and direct irritation of the eye for a week or more.

The most common complication of strabismus procedures is a failure to correct the deviation adequately. Despite exacting measurements of the angle of deviation before operation and careful measurements during surgery, individuals will vary in response to the same operation. Thus the amount of recession and/or resection attempted is based on an average response. Most patients will be corrected with one procedure, but this same procedure for the same amount of deviation will inevitably produce a certain number of overcorrections and undercorrections. Manipulation of glasses, exercises, or prisms will often suffice to complete the correction if the residual deviation is small. If not, a repeat operation may be performed on the muscles not previously operated upon, with the earlier response serving as a guide. Reoperations are common enough that they must be accepted as part of a commitment to treat crossed eyes.

Other complications of this type of surgery are quite rare. Allergic reactions to the absorbable suture material occur occasionally but are almost always easily treated with eyedrops alone. In rare cases an infection develops, or the eye is perforated inadvertently with the suture needle. Scarring sometimes limits the motion of the eye. The complications vary so much that at the most common extreme they may constitute only an annoyance, while at the other remote extreme they may result in loss of the eye. Although the odds favor an uncomplicated operation, the stakes must always be considered.

Phorias

The tropias (esotropia, exotropia, etc.) are the manifest deviations of the eyes, arising when the eyes can no longer be used together. Eyes that have a tendency to malalignment might be kept together by the fusional stimulus, the need to avoid double vision. The tendency to malalignment is called a phoria: *esophoria* if a tendency to converge, *exophoria* if a tendency to diverge.

Small phorias are present in almost everyone. It is natural to have an esophoria as a youngster, an exophoria as an adult. The transition is regular enough to give eye doctors a definite clue as to the time a given injury or disease impaired the vision of an older patient. If most of the vision is lost in an eye, it no longer has a stimulus to remain aligned with the other, for no double vision will result should it wander. Before age six it is common for an eye so affected by injury or disease to turn inward; when affected later in life, it usually turns outward.

Large phorias may cause eye strain and headache, particularly after demanding work such as reading or concentrated close work. Large phorias may be a stage in the development of an intermittent true deviation, a tropia. This sequence is particularly common with an exotropia, which usually begins as a phoria, progresses to an intermittent tropia, usually when the individual is fatigued at the end of the day, and then goes on to a constant deviation. Eye strain, headache, and double vision may become most acute as a phoria decompensates into a tropia. Another symptom of large phorias and intermittent tropias is the tendency of the patient to squint or close one eye when exposed to bright light. Parents often note this squinting, though they may be unaware of any ocular deviation in the child. The reason for this phenomenon is unknown.

Orthoptics can help by reinforcing the normal perception from each eye and exercising the ability to overcome double vision. The greater the ease with which suppression develops, the greater the likelihood the deviation will progress to a constant tropia.

Secondary Strabismus

Primary strabismus, you will recall, is caused by a perceptual dysfunction at the cerebral level which fails to keep the eyes aligned and appreciative of normal, three-dimensional vision. The exact abnormalities of the brain which cause these problems have not yet been discovered. Secondary strabismus is attributed specifically to diseases or injuries.

Strabismus may arise from organic causes such as a large retinal scar, a cataract, or corneal damage from infection or injury. When any such problem occurs in one eye, it will leave the afflicted eye at the mercy of the phoria. Since vision is so poor, no double vision results when the eye deviates. The eye that is markedly esotropic from a congenital retinal infection and the eye that turns out from an advanced cataract later in life are common examples of *organic amblyopia*. Disease or injury impaired the vision; deviation of the eye is secondary.

The preceding types of strabismus relate to sensory deficiencies. The condition may also result from disease or injury of the extraocular muscles or the nerves to these muscles. When the malalignment stems from these causes, while normal vision and perception are present, double vision will result. Only when the condition is congenital or slowly progressive will suppression intervene to avert double vision.

Direct trauma to the eye and orbit with hemorrhages, swelling, or other damage to the extraocular muscles can lead to a deviation. One fairly common injury is the "blow-out" fracture, caused by a direct blow to the eye. It is so named because the eye expands suddenly in its vertical diameter and breaks through the normally very thin bone of the floor of the orbit. The fracture site may then actually entrap the inferior rectus muscle, causing restricted motion of the eye and double vision.

Malfunction of the extraocular muscles may occur as part of a generalized disease such as myasthenia gravis, a muscle infection such as trichinosis (the pork worm), thyroid disease, and other disorders. Or the muscles may be weakened because the nerve to them has been affected. The nerves to the extraocular muscles may be injured from a stroke, a brain tumor, aneurysm, multiple sclerosis, trauma, and other problems. The sudden onset of strabismus with double vision demands immediate evaluation as it may be the first indication of a life-threatening problem.

It should be noted that true double vision, *diplopia* in medical jargon, is the perception of two distinct views, separate from each other. It is not simply a blurred image. If caused by a malalignment of the eyes, covering either eye will eliminate the second image.

The Significance of Strabismus and the Necessity of Treatment

Residual strabismus due to muscle or neurological damage or disease is treated only after full evaluation and stabilization of the condition. Because strabismus resulting from a nerve palsy—a small stroke, for example—often will recover spontaneously over a period of a year or more, surgery is not indicated before significant time has elapsed with no sign of improvement. Prism glasses may relieve symptoms in the interim. Medical treatment is available for many generalized muscle and neurological disorders, and the extraocular muscles can respond as well.

Strabismus secondary to organic amblyopia is treated for cosmetic reasons only, since no improved function can be expected from realigning an eye that does not see. However, the social stigma attached to the cross-eyed individual should not be underestimated. Large-angle deviations are immediately obvious, so individuals who are self-conscious should consider surgery.

Treatment of primary strabismus can be functional or cosmetic. If amblyopia is established and uncorrected, or if a long-standing pattern of alternating suppression is present, any attempt at realignment of the eyes must be considered strictly cosmetic. Alternating suppression, as noted before, indicates that while one eye is used, the other eye is effectively tuned out by the brain. No binocular function will be achieved with surgery.

Functional surgery is performed to get the two eyes working together again. It is most successful with the common condition of exotropia, in which patients who once had normal binocular vision slowly develop an intermittent and perhaps finally a constant exotropia. The deviation progresses as suppression of the deviating eye progresses. With realignment of the eyes, the suppression disappears and normal three-dimensional vision resumes.

Esotropias are quite different. Many children born with large-angle deviations probably lack the capacity for normal binocular vision, but early surgery has been successful with at least some of them. Success is defined as the development of some form of binocular function that will keep the eyes together and the individual aware of vision on both sides. Such rudimentary binocular vision can be termed *gross peripheral fusion.* The implication is that the individual is aware of the vision all around but lacks normal, discrete appreciation of three dimensions. A whole spectrum exists between this gross peripheral fusion and perfectly normal *stereopsis,* or three-dimensional vision. Some children have responded to early operation with nearly normal stereopsis. Exactly what constitutes an early operation, however, is still debated. Most ophthalmologists attempting a functional result like to operate before the age of four; some prefer to operate before age two.

Since the normal infant has occasional deviations of the eye prior to the development of binocular function at six months, no strabismus can be considered stable or definite before this age. Nonetheless, all large-angle constant deviations appearing before then should be evaluated by an ophthalmologist, as should any deviation persisting beyond that age. Such deviations rarely go away, despite opinions to the contrary.

Advances in anesthesia have essentially eliminated the risks that years ago were considered to be greater in very young children, and it is certainly reasonable to entertain very early surgery for functional results. Now let us consider exactly what function we hope to restore.

One-eyed individuals do very well in the real world. With experience one adapts to monocular clues to the position of objects in space. Relative sizes, shadows, superimposed objects, and already known relationships enable the individual to judge relative distances and the position of objects quite accurately with one eye alone. A child with severe amblyopia or alternating suppression has grown up adapting in this way and will suffer no obvious disability. Only when placed in an unfamiliar situation, one in which monocular clues are not available and prior experience is lacking, will the absence of stereopsis be apparent and disabling. In piloting an airplane and trying to judge the position of other aircraft in the sky, in attempting fine critical manipulations as in ophthalmology or neurosurgery, in assembling small transistors or components, for example, the lack of stereopsis may prove a real disability. In most ordinary situations, though, it will not be noticeable or disabling.

The most important treatment of a child with strabismus is avoidance of amblyopia. An individual who has lost the good eye through injury or disease will sometimes enjoy slow improvement of vision in the amblyopic eye later in life, but the recovered vision is still not normal, and such cases are quite unusual.

Reasonable attempts to gain fusion and stereopsis are justified by the functional significance of preparing a child for a visually demanding society, and by the cosmetic significance of avoiding a potentially embarrassing and psychologically debilitating disfigurement.

8 Retinal Detachment, Retinal and Vitreous Disorders, Color Blindness

Retinal detachment is a relatively common problem which can cause total blindness in the eye involved. The retina is the tissue paper-thin inner lining of the back part of the eyeball. It is like the film in a camera; the perfectly arranged cells of the retina register the pattern of light that is focused upon it and then transmit it to the brain. Any problem affecting the retina thus affects vision directly. This tissue-thin lining of the eye is attached to the outer coats of the eye only at the front end (the anterior insertion) and at the optic nerve. Figure 3–2 demonstrates the position and normal attachment of the retina. Filling the center of the posterior portion of the eye (called the *vitreous cavity*) is a clear gel (the vitreous) and a variable amount of fluid. If a hole or tear develops in the retina, the fluid from the center of the eye may flow into the hole and get under the retina, causing it to separate from the outer coats of the eye. This begins the process of retinal separation, or *detachment*. It almost always progresses until the retina has detached totally into a folded funnel-like configuration about the optic nerve. At this point the eye is blind. A detached retina cannot function because the outermost cells of the retina depend upon the blood vessels of the choroid, the intermediate layer of the eye upon which the retina normally lies. Deprived of the nutrition and oxygen from the choroid, the outer retinal cells gradually die.

There are several different types of retinal tears and holes. Some holes develop as degenerative changes in older individuals. Reduced blood supply to areas of the retina is one of the aging processes that can lead to focal retinal thinning and finally hole formation. Holes also form in cases of extreme nearsightedness, associated with an increased size of the eyeball. The retina is actually stretched and thinned, thus promoting hole formation. Approximately 7 percent of the population suffer from a congenital condition known as *lattice degeneration of the retina,* in which focal bands of thinning and degeneration lead to hole formation. Less common congenital conditions also result in hole formation due to retinal thinning or degenerative change.

In addition to abnormal thinning and degeneration of the retina, the vitreous gel in the center of the eye may contribute to hole formation by abnormally firm focal attachments to the retina. With sudden motion of the head, this gel may then pull on the retina and tear it. Nonetheless, trauma has not been definitely shown to be a cause of most retinal tears and detachment. Almost everyone can recall an instance in which the head was suddenly jarred or the eye was poked. The extent to which these seemingly minor events really cause retinal tears is quite conjectural. Of course, severe ocular trauma, especially injuries in which the vitreous gel is actually penetrated or some vitreous is lost through a penetration site, will usually scar the vitreous and create traction on the retina, causing detachment.

Infections within the vitreous cavity of the eye, advanced proliferative diabetic retinopathy, and other proliferative retinopathies (in which abnormal blood vessels and scar tissue grow out of the retina into the vitreous) also lead to vitreous scarring, traction on the retina, and, finally, detachment. Rarely, severe inflammation within the choroid or retina, tumor, or high blood pressure will cause an *inflammatory* or *serous* detachment from accumulation under the retina of fluid which has leaked out of abnormal blood vessels or through an impaired barrier between the retina and choroid. The most common type of retinal detachment, however, is that associated with a tear or hole in the retina. Figure 8–1 is a drawing of a typical retinal tear in a partial retinal detachment. Figure 8–2 represents an end-stage total retinal detachment.

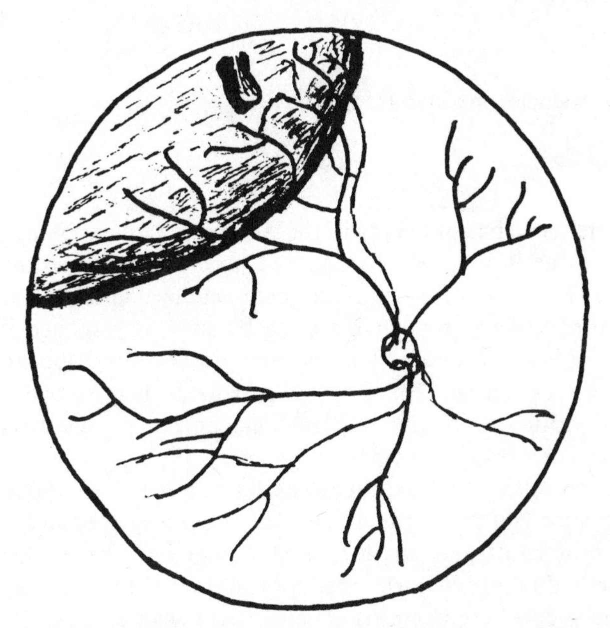

Fig. 8-1. A view inside the posterior eye demonstrating a tear in the retina at the 11:30 position and a partial retinal detachment caused by it.

Fig. 8-2. A total retinal detachment in which bulbous folds of retina have peeled away from the outer coats of the eye and almost touch in the center.

Retinal Detachments and Retinal Tears

Incidence

Retinal detachment occurs in the American population at a rate of about one in nine or ten thousand per year. One percent to 3 percent of the eyes operated on for cataract develop retinal detachment after surgery, though newer techniques may reduce its incidence. Localized and early symptoms do not often appear after a cataract operation. The first thing patients may notice is decreased vision, but by this time the detachment is quite extensive, involving the macula and posterior pole.

The incidence of retinal holes and tears, however, is quite high. Approximately 9 percent to 11 percent of autopsy eyes and a similar percentage of patients in clinical studies have been shown to have retinal breaks. Clearly, then, not all retinal breaks will result in detachment. Certain coexistent conditions aside from an obvious early detachment about the break guide the ophthalmologist in deciding which holes might lead to a detachment and deserve treatment. Extreme nearsightedness, an eye operated on for cataract, and a history of a detachment in the opposite eye suggest that the tear will progress to detachment and should be treated.

Symptoms of Retinal Detachment

Once a tear or hole has formed and fluid begins to seep through it and under the retina, the retina starts to peel away. As almost all of the holes and tears develop anteriorly in the periphery of the retina, where the retina is thinner and the vitreous commonly has attachments, the detachment will begin peripherally. It then extends backwards toward the posterior pole, which represents the keen central portion of the visual field. Initially it may seem as if a curtain or veil is being dropped into one's field of vision. If it progresses from below, it will seem as if a curtain is coming down from above (recall that the projection of the vision is upside down and backward); if it starts above and peels down, it will seem as if this curtain is being drawn up from below (figure 8–3). Similarly, if it progresses from one side, it will appear to approach from the opposite. Only if the detached retina drapes over the

Fig. 8-3. Scene above would appear to an individual with a partial retinal detachment, as drawn in Fig. 8-1, as if a curtain shaded the lower corner of the field of vision of that eye (below).

macula, or if the detachment extends posteriorly to the macula, will one's vision straight ahead be impaired.

Symptoms of Retinal Tears

Like actual light rays, pressure on the retina will cause the retinal receptor cells to fire. Thus, when one is hit in the eye or pushes a finger against it, momentary flashes of light will appear. Light is also seen when a vitreous strand attached to the retina exerts traction upon it in the process of forming a tear; or when the retinal cells are stimulated as the retina peels away. If a hemorrhage is associated with the tear formation, as it will be if a small retinal vessel crosses the tear, many black spots, a fountain formation, or odd inkblot shapes will drift around in one's vision. It may progress so that blood fills the vitreous cavity and only light can be seen. When a large vitreous hemorrhage is present, it may be difficult for the ophthalmologist to diagnose a retinal tear or detachment without the use of special techniques. Just as the blood prevents the patient from seeing, the ophthalmologist's view of the retina may be obscured as well.

Treatment of Retinal Detachment

From the previous discussion, it is apparent that retinal detachment occurs when one or more retinal holes allow fluid to enter the space beneath the retina and cause it to peel away. Treatment of a detachment will try to seal any such holes in the retina—no small task when the retina and the tears are already detached and floating out in the vitreous cavity. A laser beam will shoot right through the retina and effect no adhesion between retina and choroid, as they are too far apart. For the same reason, applications of cold (*cryotherapy*) or heat (*diathermy*) in an effort to inflame the choroid and get the retina to stick to it do not usually work. Thus, a technique called *scleral buckling* was developed. The eyeball in the area of the retinal tear is indented so that the choroid and retina move closer together and seal the hole. There are a variety of techniques for making this indentation and creating a buckle, but most utilize inert silicone sponges or bands. The indentation relieves traction on the retina from vitreous strands.

The fluid that has accumulated beneath the retina may be drained out through a carefully placed puncture site, if it is necessary to do so to close the retinal hole or hasten recovery.

Retinal detachment procedures are done under local or general anesthesia, depending on the surgeon's preference, patient's condition, and anticipated complexity of the case. After the procedure, both eyes may be patched to keep them as still as possible while the retina is allowed to settle. Complete bed rest with or without both eyes patched may be necessary for an indefinite period of time after the operation, depending on how readily the retina settles into place. The patient may be up and about right away if the retina is flat and the tear well sealed right after the operation. The postoperative course thus varies with the type and extent of detachment and the response to surgery.

Treatment of Retinal Breaks

Treatment of a retinal hole with minimal or no surrounding detachment is a relatively simple procedure. The aim of treatment is to seal the break so that no fluid from the vitreous cavity can flow through it and get under the retina to cause detachment. The ophthalmologist produces focal inflammation, which causes a small scar and thus adhesion between the retina surrounding the tear and the choroid. Presently, the most common ways of producing such focal inflammation are the laser and the cryoprobe. If the laser is used, concentrated light energy burns the retina and choroid. The quick and painless laser treatment is usually done at a slit lamp with a special contact lens placed on the eye. Depending on the location of the tear, cryotherapy may be applied with equal simplicity. This form of treatment employs a freezing probe that is placed over the site of the tear once the eye has been numbed with appropriate medication. It is activated momentarily and the area about the tear frozen. If the tear is too far posterior, it may be necessary to make a tiny incision in the conjuctiva to get the probe in the proper position. Even if an incision is necessary, the treatment causes little or no discomfort and complications are rare.

If a retinal tear is quite large or has detachment around it, it may be best to treat it as a retinal detachment.

Complications from the Treatment of Rentinal Tears and Detachment

The treatment of retinal tears with cryotherapy or laser is almost always uneventful. Very rarely will an eye react with an unusual amount of inflammation that will cause excessive scarring of the retina or vitreous. The scarring can then lead to distortion of vision or even retinal detachment. Infection or hemorrhage may occur, either as a mild, transient problem or a devastating, permanently damaging complication. The development of retinal detachment from a tear, however, is more likely to cause loss of vision than any complication of treatment.

As a retinal detachment procedure is more complicated and extensive than the treatment of a tear alone, it has a higher rate of complication. In addition to those rare problems encountered after treatment of tears, detachment procedures may result in transient or permanent muscle imbalance, glaucoma, cataract, and persistent inflammation in the eye. Despite these potential complications, the overall success rate for retinal detachment procedures is 90 percent or higher. This degree of success is impressive, considering that only thirty years ago, before these techniques were developed and perfected, blindness was the inevitable result of retinal detachment.

Prognosis for Vision after a Retinal Detachment Procedure

Successful surgery does not always restore vision to the level it was at before detachment. The recovery of visual function depends on the extent of irreversible damage the retina has undergone while detached. If the detachment is peripheral and the macula and central retina were not detached, there will be little or no obvious visual impairment after uncomplicated reattachment. Once the central vision has been affected by detachment of the corresponding part of the retina, residual defective vision may occur despite successful reattachment. Such factors as duration of the detachment prior to operation, age of the patient, and general status of the retina affect the outcome by determining how much irreversible damage occurred during the period of detachment. The ophthalmologist can usually give a reasonable idea of what to expect, though the damage that has occurred during detachment is at a cellular level and hence inconspicuous. If symptoms of visual loss are immediately apparent to the patient and he receives prompt treatment, normal or almost normal vision can be restored in most uncomplicated cases.

Traction Detachments

Fortunately, most retinal detachments are caused by retinal tears and are amenable to the operative procedures mentioned previously. Some, however, are caused by scarring of the vitreous gel in the center of the eye. These vitreous strands and scars attach to the retina and contract, actually pulling the retina off. With extensive detachments from this cause, the routine detachment procedures are often ineffective. Such vitreous scarring can result from advanced diabetic retinopathy, severe injuries, vitreous hemorrhages, and other less common disorders. One of the most exciting recent developments in ophthalmology over the past decade has been the improvement and perfection of instruments capable of removing vitreous scarring. They relieve the traction on the retina that has caused the detachment so that reattachment can occur. These instruments are probes that can be inserted into the vitreous cavity. Called *victrectomy* instruments, they allow the doctor to cut and aspirate scars and other material while maintaining an infusion of fluid to keep the normal shape of the eyeball.

Vitreous Hemorrhage

Hemorrhages can occur in the vitreous cavity from hardened arteriosclerotic blood vessels, from diabetes and other abnormalities of the blood vessels, or from a retinal tear resulting in a ruptured vessel, or from a tumor. If they are extensive, the hemorrhages cause marked loss of vision. Vitrectomy instruments have also been used to remove the residue from vitreous hemorrhage. Since the operation has significant complications, however, adequate time is usually given for possible spontaneous clearing of the hemorrhage. Clearing may extend over a year or two after the initial bleeding episode. Special techniques and ultrasound examinations can be used to assure that a retinal detachment or tumor is not being masked by the hemorrhage.

Vitreous Floaters and Vitreous Detachment

Everyone has tiny imperfections in the vitreous gel, present at birth and increasing somewhat with age. *Vitreous floaters,* a few small black lines or spots that are noted to drift across one's field of vision when looking at a plain background or while reading, are almost never signs of a hemorrhage. As one ages, the vitreous gel tends to liquefy and collapse. This process of vitreous detachment causes some traction on the retina and will result in spontaneous flashes of light as well. Vitreous detachment and floaters, however, are normal accompaniments of age. Vitreous detachment occurs in up to 50 percent of individuals by age fifty and about 90 percent of individuals by age seventy. It occurs earlier and more commonly in nearsighted people. The symptoms of vitreous detachment and retinal detachment are similar, but the two conditions can be differentiated by an ophthalmologist, who dilates the pupil and examines the vitreous and retina directly.

Vitreous Floaters and Their Significance

The early symptoms of retinal detachment or retinal tear formation are similar or identical to those of vitreous detachment. A small vitreous hemorrhage or a vitreous detachment can cause the same symptoms of spontaneous light flashes or floaters. The conditions at this stage can be differentiated only by the ophthalmologist, who examines the retina and vitreous through a dilated pupil. Other retinal conditions may exhibit the same symptoms, but one must bear in mind that vitreous detachment and deterioration is an inevitable accompaniment of aging. It is wise to have the eyes examined when floaters first appear, but there is no need for alarm or reexamination each time small floaters reappear or change in configuration. The normal floaters are small dots, specks, and curvilinear black veil-like opacities noted most often when looking at a plain background such as the sky or a white ceiling. Changing the eye's focus or looking at a busy or patterned background will cause them to fade from view. As the eyes move, these floaters usually lag behind a bit. They can be most annoying when they drift into the center of vision while one is reading, but they can be moved out of view by moving the eyes. They can be noted sometimes while looking toward a light with the eyes closed. When one's attention is directed to these floaters, they can be seen, but if one is assured that they are normal and then ignores them, they will assume their rightful place as an occasional nuisance only. A sudden shower of floaters or a large, expanding inkblot floater, however, may indicate a hemorrhage in the vitreous. Any associated visual loss, central or peripheral, deserves immediate attention, as a retinal detachment may be developing.

The ophthalmologist will put drops in the eyes to dilate the pupils and then with special instruments examine the retina and vitreous. If only vitreous detachment or vitreous floaters are present, routine periodic examinations will suffice, and the patient should be reassured that the condition is normal. No medicines or eyedrops will eliminate the floaters, and certainly no operation is indicated to remove them. If the floaters are ignored, they will disappear from consciousness over time.

In summary, the prognosis for the vast majority of patients with retinal detachment is now excellent due to the great strides in diagnosis and therapy made over the past several decades. Early diagnosis is the key to successful treatment. Prophylactic treatment of suspicious retinal tears is often preferred even before detachment occurs. Thus, while floaters and flashes of light are almost always symptoms of normal vitreous processes, a complete ophthalmologic evaluation is indicated at the onset of such symptoms in order to be certain there are no retinal problems that are best treated early.

Retinitis Pigmentosa

Retinitis pigmentosa is an inherited disorder that causes slow deterioration in the retina, the "film" in the eye that perceives light and is thus necessary for vision. Clumps of pigment develop in the retina, the retinal blood vessels become narrowed, and the optic nerve atrophies. The first and most prominent symptom of this disease is night blindness. The symptom must be put into perspective, though, for everyone sees less keenly at night and in the dark. The retina normally shifts from very keen acuity in the light, using mostly cone cells, to less discriminating but more highly light-sensitive vision in the dark, utilizing the rod cells.

Thus, we see better in the dark by looking slightly off to the side of an object. The central fixation point and source of our keenest vision is the fovea, composed exclusively of cones. Because only the more light-sensitive rods function in the dark, we cannot discern as much detail under these circumstances. Retinitis pigmentosa sufferers often cannot function at all in dimly lit environments that are adequate for normally sighted individuals.

Retinitis pigmentosa has been found as a sporadic disorder in individuals and as an inherited disorder in families, the pattern of heredity varying from family to family. It has been seen in all forms: dominant, recessive, and sex-linked. It is rarely part of one of numerous syndromes characterized by a constellation of inherited abnormalities. The severity of the retinitis pigmentosa is fairly consistent in any given family or syndrome. It may be stable, slowly progressive, or it may lead rapidly to total blindness in the second or third decade of life. It can begin with gradual loss of the peripheral field of vision, progress to formation of cataracts, and culminate in complete deterioration of the retina and optic nerve.

Experimental evidence suggests that a retina afflicted with this condition may preserve its function longer if it is shielded from light. Doctors are testing opaque contact lenses, applied to one eye. Military sighting devices for night vision are being adapted for retinitis pigmentosa patients. The most promising is a pocketscope that greatly enhances the level of illumination. Since the causes of retinitis pigmentosa are not yet known, no effective therapy is available.

Color Blindness

True color blindness, in which all colors appear only as shades of gray, is extraordinarily rare. Less severe forms of defective color vision, however, are quite common, existing in 8 percent of the male and 0.5 percent of the female population. The reason for the preponderance of defective color vision in males is that these common forms are inherited as *sex-linked recessive disorders.*

The genes determining these forms of color vision are located on the X chromosome. Since males have only one X chromosome, they will exhibit a trait localized on this chromosome. Females have two X chromosomes, so one of the two can be affected without exhibiting the trait. Females who have one affected X chromosome are called *carriers* and statistically will pass defective color vision on to half of their male offspring and cause half of their female offspring to be carriers. Both X chromosomes must be affected before a female will exhibit defective color vision.

The most common form of defective color vision, found in 5 percent of all males, is called *deuteranomaly*. It affects one's ability to distinguish green and yellow. Confusion also exists with purple, azure, green, gray, and other mixed hues containing greens. *Deuteranopia,* a more severe form of deuteranomaly found in one percent of all males, causes even more profound problems with hue distinctions. *Protanomaly* and *protanopia* are each found in about one percent of the male population and represent a similar spectrum of defective color perception, involving primarily reds, yellows, and mixtures of blue-green and gray. Typically, red, gray, and blue-green are confused.

The functional significance of these defects varies with their severity and the occupational needs of the individual. Most people with deuteranomaly are not aware of their problem until faced with a testing situation, whereas protanopes notice a marked decrease in their sensitivity to reds. They can learn to identify the colors by association and under normal circumstances experience no real disability. Of course, jobs requiring critical color discrimination may exclude individuals with significant defects. Little can be done medically to give these people better color discrimination for their occupations. A tinted contact lens will help somewhat, but because it confuses the colors that are perceived normally, it can be worn on one eye only.

9
Macular Degeneration; Optical Aids

Approximately 65 percent of the cases of diminished vision in the United States are due to macular degeneration. The disorder is common in elderly people. It does not cause total blindness but results in loss of central vision, which in advanced cases can be very disabling. Since no medical treatment for the problem has proven uniformly effective, optical aids such as magnifiers and telescopes are used to improve visual function. This chapter will explain the nature of the common forms of macular disease, the rationale for evaluation and treatment, and the types of optical aids available when medical treatment is inappropriate or ineffective.

Function of the Macula

The macula is the central area of the retina directly behind the pupil. It receives the most critically focused central light rays and thus provides the central portion of the visual field. It has the highest concentration of the keenest receptor cells (cones), arranged in a perfectly regular fashion. These cells enable the eye to appreciate fine detail as well as color. Cones are scattered about the retina in other areas, but it is only in the tiny macular area that their number and arrangement is sufficient to allow good visual acuity. Cone cells provide color vision as well as keen acuity. Thus, with disorders of the macula, color perception as well as acuity may be impaired.

The entire macular area measures approximately 3 millimeters in diameter. If this area does not function properly, the vision is reduced to the level of 20/400 or less; that is, one can see only the large E at the top of the typical vision chart, or less. The central 0.1 millimeter, called the foveola, is necessary for normal keen 20/20 vision; so one can appreciate the critical function of this very tiny area, literally the size of a pinhead. The central vision is variably diminished depending on the location and extent of the nonfunctioning cells caused by macular disease. Figure 9–1 is a diagram of the macular area of the retina.

Symptoms of Macular Disease

Loss of central vision is the main symptom of macular disease. The patient with advanced macular degeneration must look off to the side in order to bring an object into view. As mentioned before, it is only in the central macular area that the keen visual receptor cone cells are in high enough concentration to provide good vision. Thus, objects are not seen clearly even when brought into view by using the more peripheral retina. The exact size of the central area in which one does not see depends on the size of the area in the macula that is scarred or not functioning. Figure 9–2 shows how parts of a scene become obliterated by progressive loss of central vision.

Less advanced cases of macular disease are not so disabling. Minimal deterioration of vision usually occurs without such an obvious central *scotoma* (nonseeing area). Some loss of color discrimination or distortion of central vision may become obvious. The distortion is seen as bulging or curving of straight lines or contours. Doorways, for example, may seem buckled. It indicates a corresponding distortion of the array of retinal receptor cells in the macula. The extent of such symptoms will vary with the extent of the disease process. The onset may be sudden or very gradual and almost imperceptible, again depending on the nature of the particular process disturbing the function of the macular cells.

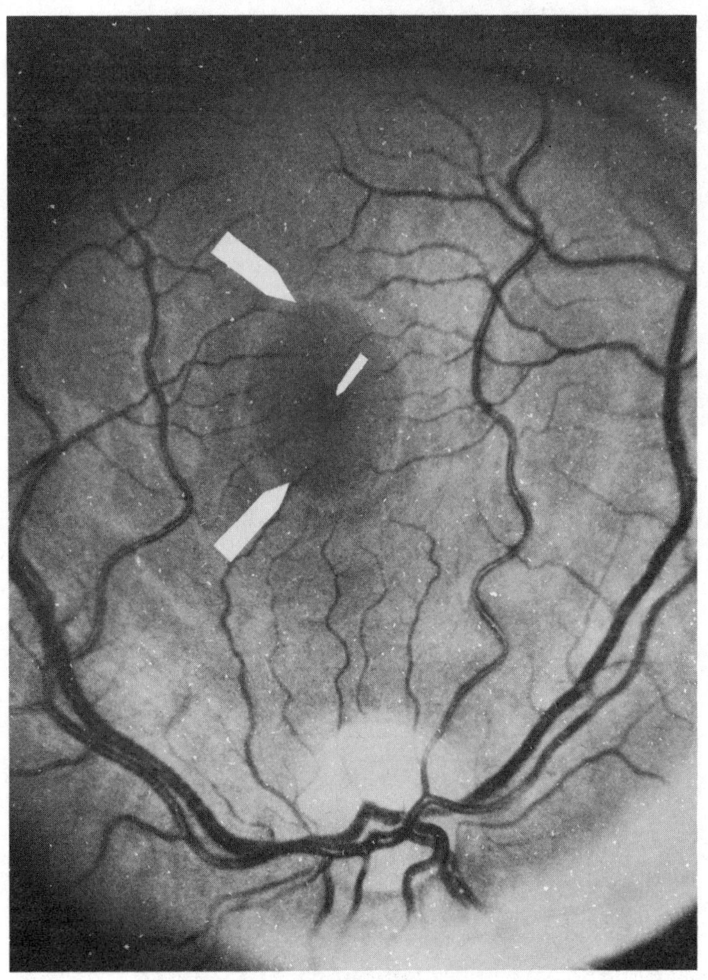

Fig. 9-1. The posterior pole of the eye demonstrating the macular area (between larger arrows), fovea (small arrow), and optic nerve.

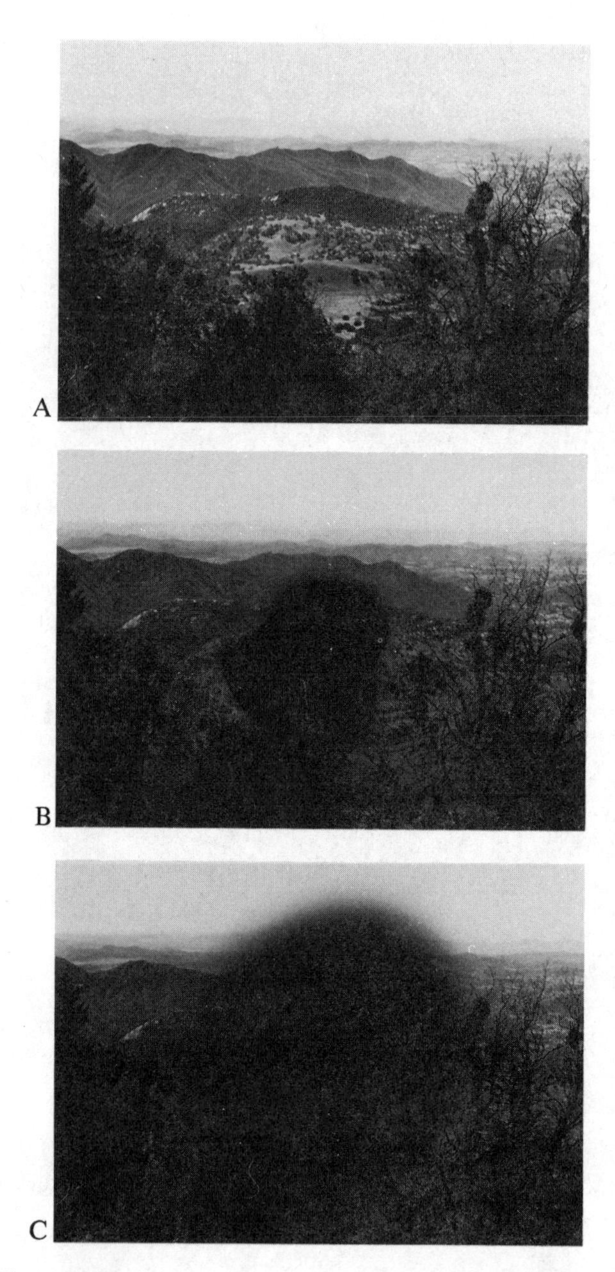

Fig. 9-2. The lower photographs (B and C) illustrate varying degrees of loss of central vision typical of macular degeneration.

Types of Macular Disease

Loss of macular function may occur from a variety of processes, all of which produce one or more of the symptoms just described. There are several congenital and hereditary macular degenerations that can occur during youth and early adulthood. Blood-borne infections that are carried into the eye may affect the macula, as can many other diseases, infections, and trauma. If the macular problem is due to an active infectious process or a tumor, appropriate therapy is applied as soon as an accurate diagnosis is made. Most commonly, however, loss of macular function occurs as a consequence of poorly understood degenerative changes associated with aging. When degenerative processes or old inactive scars are present, the therapeutic alternatives are limited.

In all such disorders of the macula one or more common factors can be the cause of loss of function. The visual receptor cells themselves in the macular area may simply degenerate, due either to poor blood circulation in the area or primary (perhaps hereditary) degeneration of the cells themselves. Blood vessels that supply the macular area, either from the retinal or the choroidal circulation, may be afflicted with disease processes such as arteriosclerosis (hardening of the arteries), inflammation, infection, high blood pressure, or diabetes. If the blood vessels leak, excess fluid and lipids will accumulate in the retina; if they break, small or large hemorrhages may occur. Scar tissue may form from any of these abnormalities. The abnormal new blood vessels that develop in the scar tissue are fragile and may leak and hemorrhage easily.

Normally, a selective barrier between the retina and choroid allows only the proper amount of fluid and nutrients from the choroidal circulation to pass over to the outer retinal layer. This barrier consists of a fine membrane and active pigment cells, called the pigment epithelium. Degenerative, inflammatory, or infectious processes may produce defects in this barrier and allow an excess amount of fluid to pass into the macular area. If scar tissue forms under the pigment epithelium, it will lead to abnormal fragile blood vessels that can hemorrhage and create even more scarring. Once definite scar tissue is present in the macula, visual function in that area is lost. Hemorrhage will reduce vision and sometimes block it totally, though it may return when the hemorrhage clears. Excess fluid and swelling will distort the array of visual cells and provoke a similar distortion and reduction in vision.

Diagnosis and Therapy

Diagnostic techniques are directed toward differentiating these processes. Treatment halts leakage through the pigment epithelial barrier before irreversible changes have occurred, and it seeks to obliterate abnormal blood vessels before hemorrhaging and scarring progress. Leaks can be sealed and vessels obliterated with laser therapy, but not until the exact sites have been identified. The laser is a very powerful light source that produces discrete burns in the eye. The small scar resulting from the burn will seal a leak or occlude a small vessel. But a scar does result, so it is pointless to treat exactly in the center of the macula, for one then destroys the cells necessary for recovery of vision. The extent and location of the abnormalities must be carefully evaluated and the risks of therapy considered. If the abnormal sites are not directly in the fovea, or central one millimeter, therapy may be possible. Of course, laser therapy will not improve the vision lost from already established scars, occluded blood vessels, or degenerated, irreversibly damaged visual cells.

Fluorescein Angiography

One of the most valuable techniques for the evaluation of macular disease is fluorescein angiography. A small amount of yellow fluorescein dye injected into a vein in the arm or hand travels throughout the bloodstream. A series of photographs is taken in rapid sequence as this dye passes through the blood vessels of the choroid and retina. An ultraviolet filter is placed in the camera (or in an ophthalmoscope) so that the light transmitted into the eye will cause the dye to fluoresce and allow observation of processes otherwise hidden to view. The dye shows up in the blood vessels and leaks out of abnormalities along with fluid from the bloodstream. Abnormal vessels, hemorrhages, leakage sites, and scar tissue can be visualized with this technique. Figures 9–3 through 9–9 show different types of macular degeneration and angiographic findings. These findings are correlated with the findings

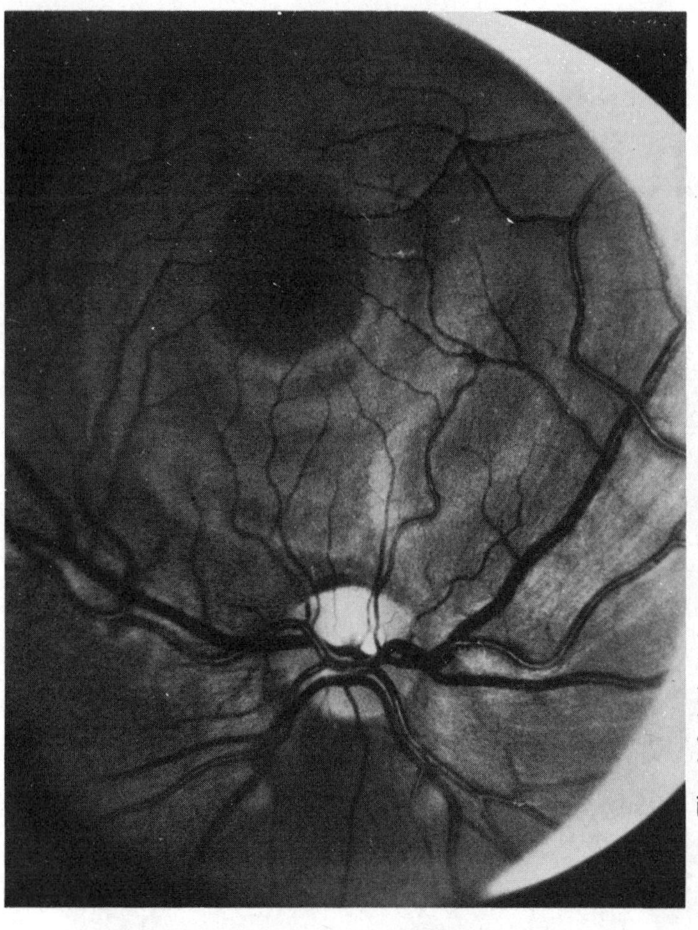

Fig. 9-3 (A). A normal macular area as it appears during fluorescein angiography.

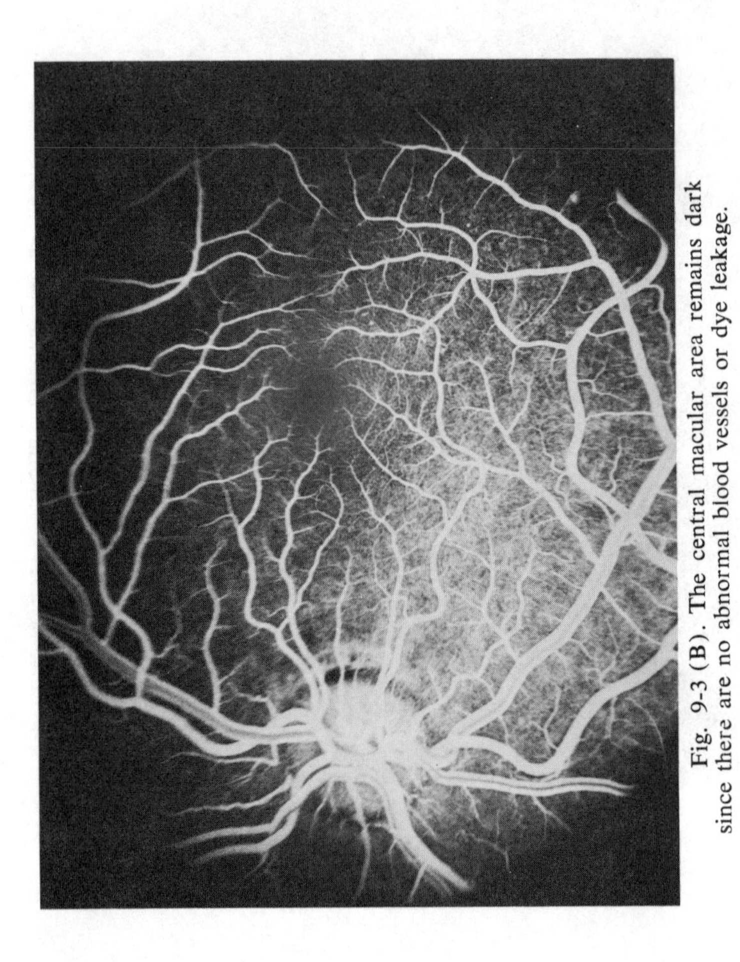

Fig. 9-3 (B). The central macular area remains dark since there are no abnormal blood vessels or dye leakage.

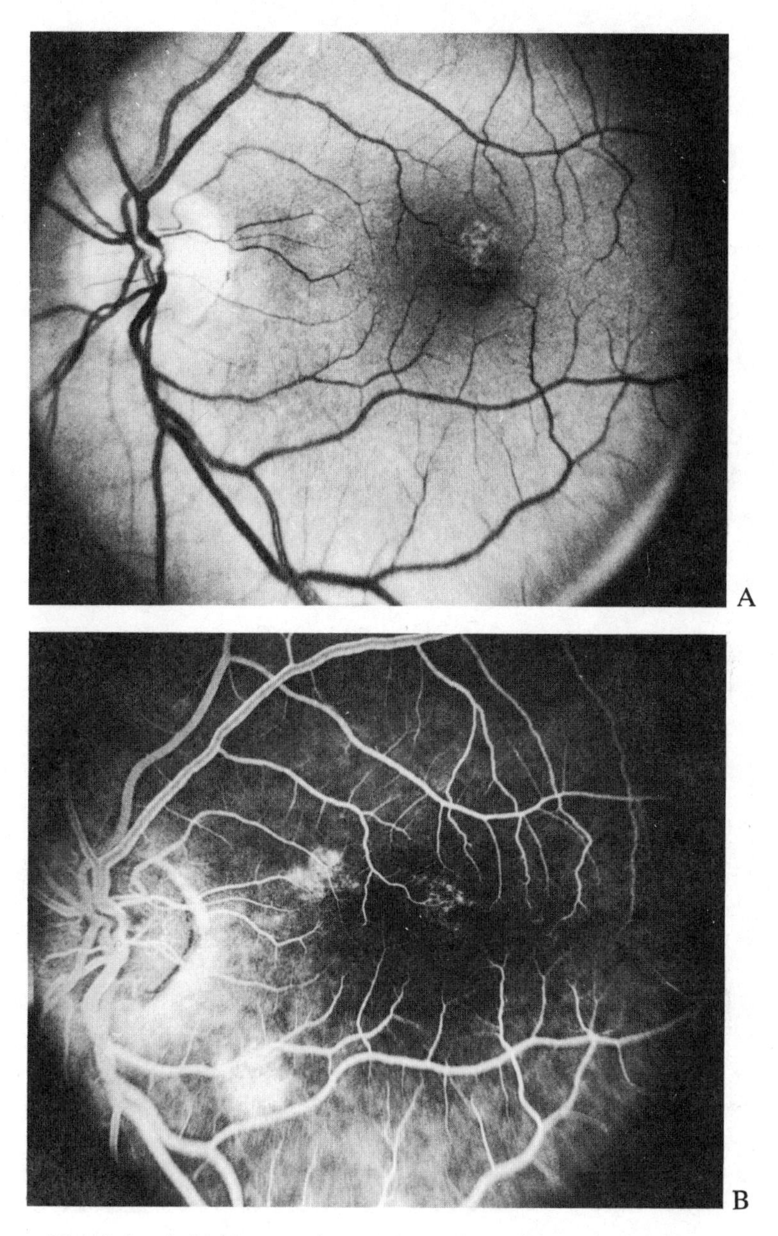

Fig. 9-4. A focal area of loss of the outer pigmented retinal layer (A) is highlighted in the fluorescein angiogram (B), as are other areas of degeneration.

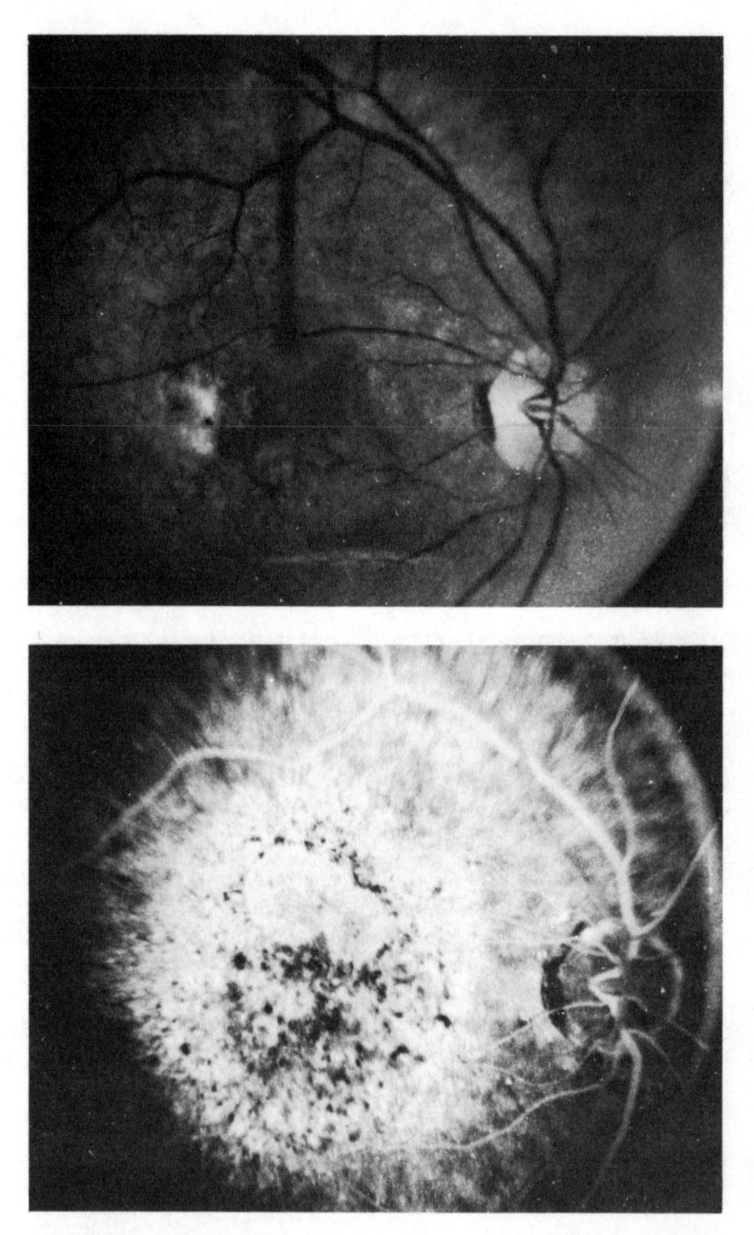

Fig. 9-5. More profound and diffuse abnormalities of the outer retinal layer in the macular area are highlighted in the fluorescein angiogram (**B**).

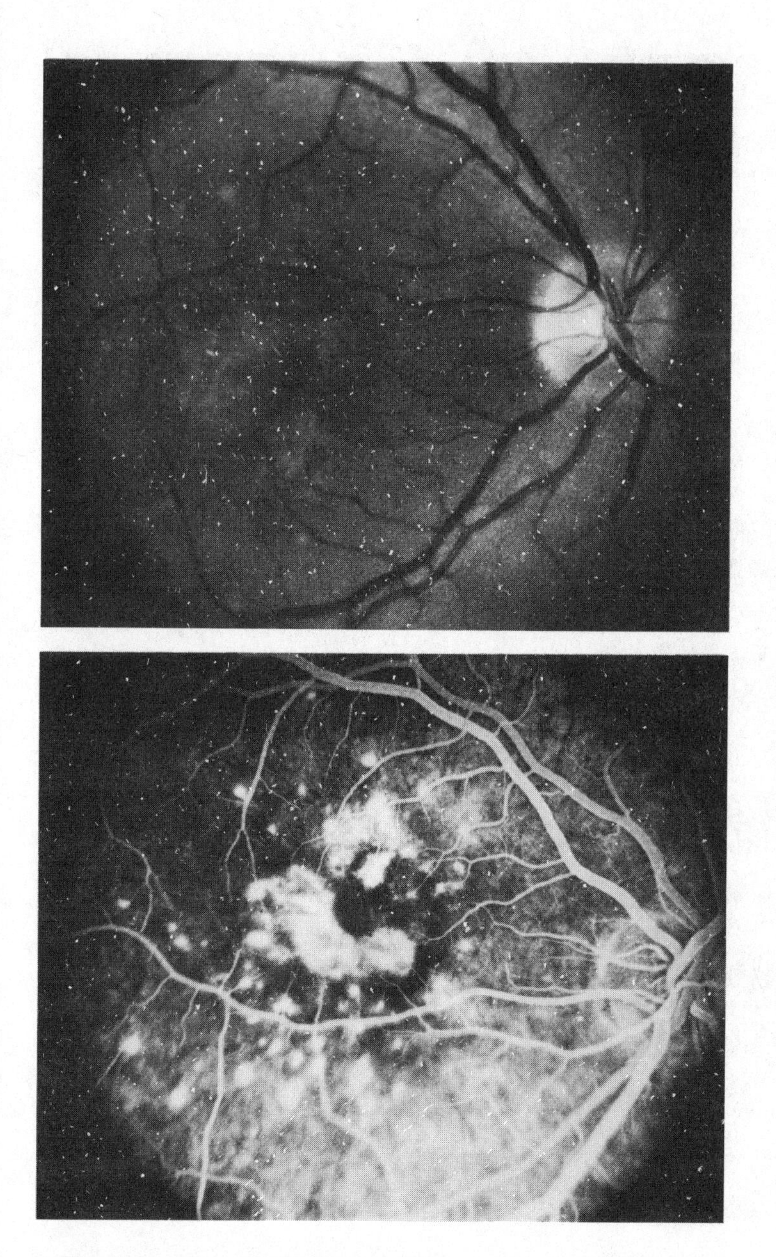

Fig. 9-6. Abnormal blood vessels have grown from the underlying choroid and have caused some hemorrhage deep in the retina.

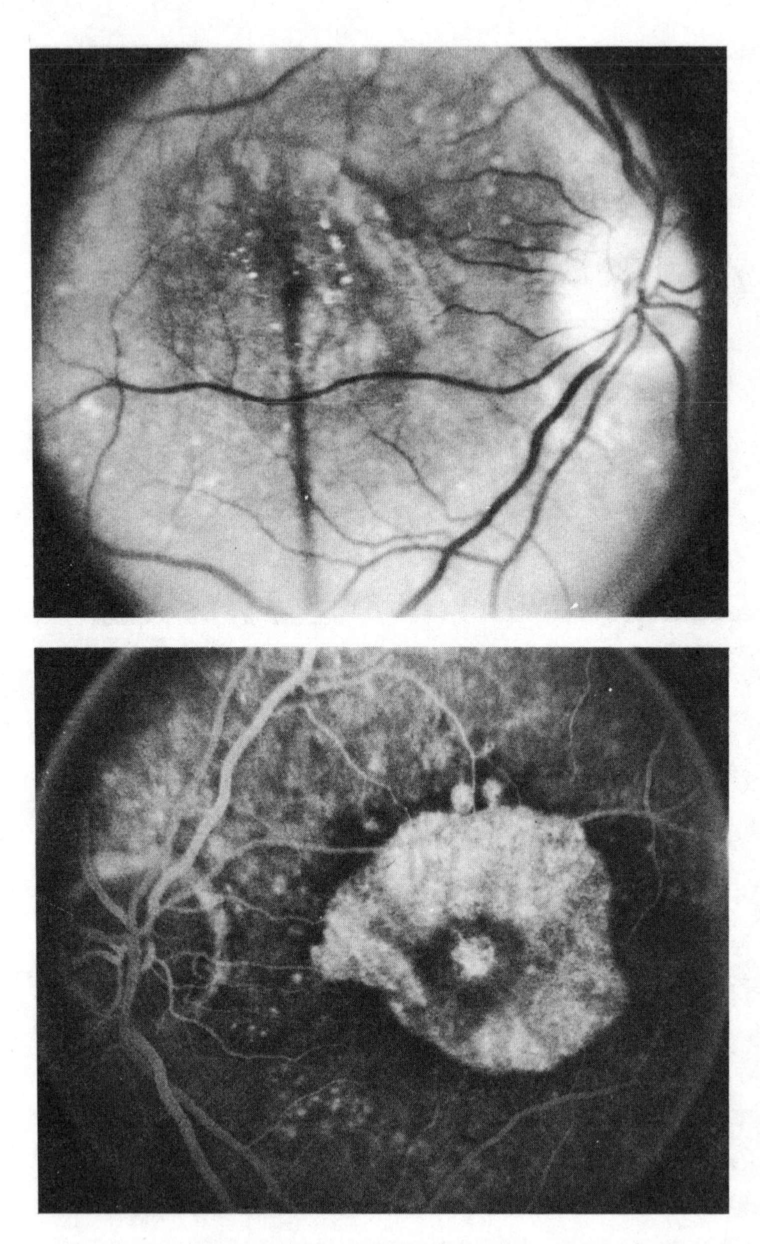

Fig. 9-7. Diffuse leakage of fluid from abnormalities in the macular area is seen as dye leakage.

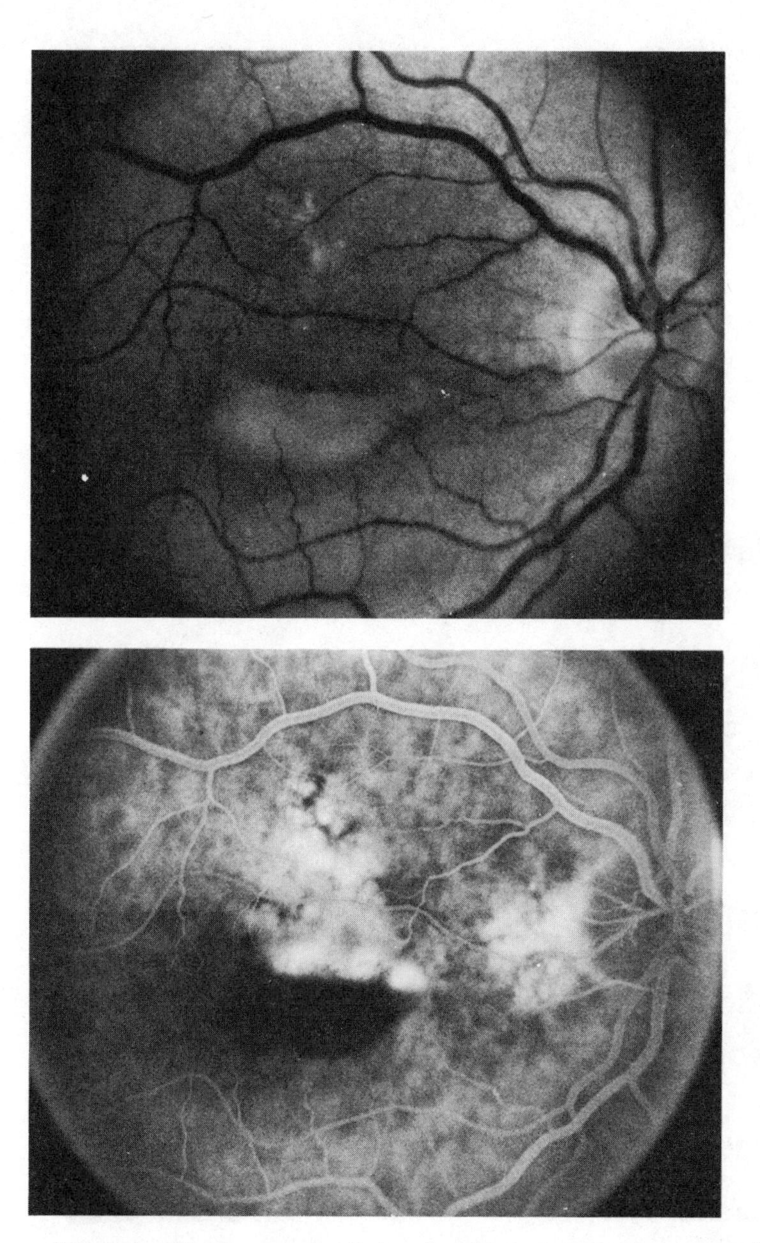

Fig. 9-8. Fibrous scar formation is taking place in the areas of hemorrhage and leakage. The scar appears dark in the angiogram because dye is not seen through it.

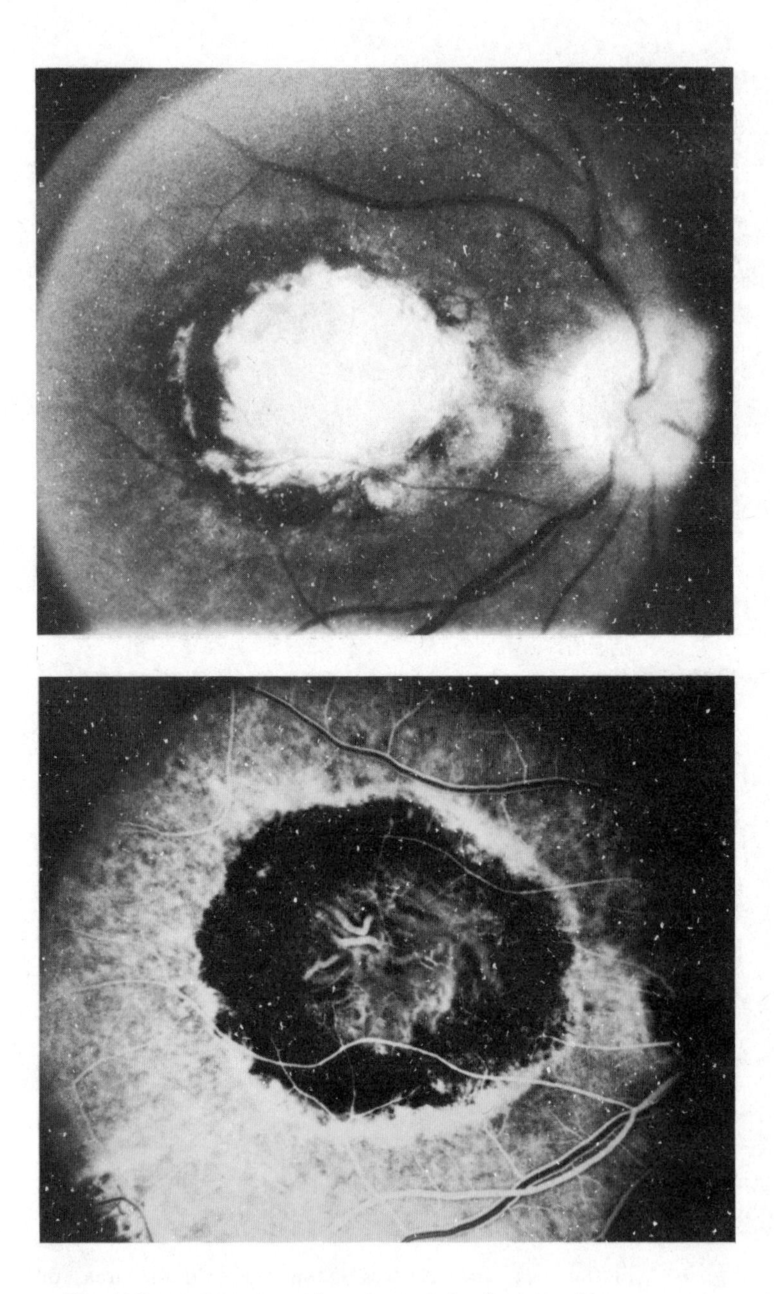

Fig. 9-9. A large scar has formed in the macular area over profound degenerative changes and abnormal blood vessels.

on examination. The ophthalmologist utilizes ophthalmoscopes, diagnostic contact lenses, and other instruments to provide a highly magnified, three-dimensional view of the macula.

To date, there is no definitive, well-controlled study of the effects of laser treatment on macular degeneration. Studies have been hampered by the long follow-up necessary to determine whether useful vision is indeed preserved longer in a treated than in an untreated eye and by the tremendous variability in the untreated natural course of the disease. Most ophthalmologists agree that laser treatment has been disappointing in most cases. A few people respond dramatically with improved vision, others maintain the existing level of vision, and many worsen just as rapidly as they would otherwise. Vision sometimes deteriorates even more rapidly, perhaps due to laser treatment, which may occasionally precipitate a hemorrhage or more scarring. Before undergoing treatment, then, one must try to determine whether vision will worsen if left untreated or improve on its own, without treatment.

Not all macular degeneration gets progressively worse. When due to aging and degenerative processes, it is usually more or less symmetrical at the end, inactive stage. But striking exceptions occur, especially in the rate of its development. One eye may precede the other by many years. The course of the first eye involved is usually, but not always, a guide to the course of development in the other. However, there is so much variability from one patient to the next that it is virtually impossible to predict with certainty exactly how bad vision ultimately will be when the degeneration stabilizes.

The good news in this discussion is that macular degeneration never results in total blindness. Most commonly, it is noted by the ophthalmologist as a minimal diminution in visual acuity that is unnoticed or only slightly apparent to the patient. Little or no progression occurs. At its worst, a large central scar or area of degeneration may develop. When this happens, the patient will see nothing straight ahead and will be aware of a large central dark area of nonseeing in the visual field. Looking off to the side will allow a glimpse of objects that are straight ahead. The size of this central blind area varies from case to case; the larger it is, the worse the visual acuity will be. But peripheral vision will remain in even the most advanced cases, and patients almost always will be able to get around by themselves. The greatest losses are in reading and keen vision. The solution to this problem will be considered next in the discussion of optical aids.

Optical Aids

Optical aids help partially sighted individuals adapt to their handicap and continue as normal a routine as possible. These devices augment all types of subnormal vision. They are mentioned in this chapter because the majority of individuals seeking such help in low-vision clinics throughout the United States have macular degeneration. Patients with diminished vision from diabetes or other retinopathies, cataracts, glaucoma, neurological diseases, and other problems find these aids helpful as well.

Adequate lighting is important to those with diminished vision. It is not always a matter of more intense light, however. With certain diseases, glare becomes a tremendous problem. The time spent experimenting with various types of light and positioning the light source to the side or over one's shoulder to be directed onto the working or reading surface will pay off in increased comfort and ease of vision. In general, gooseneck or flexible lamps are the most versatile. If positioning for maximum light intensity causes too much glare, visors should be considered.

Large-print books and periodicals have been a great boon to people with subnormal vision. Chapter 17 discusses these publications and where to find them. Jumbo playing cards restore the joy of card playing for those who can no longer see the regular cards or are too embarrassed to use optical aids during a game. Large-number telephone dials are often given away as promotional devices by local stores. The numbers are usually raised and can be easily felt if not seen. Also helpful are line guides—cutout cards that isolate a given line for reading or have cutouts set up to correspond to lines on a check or form. The individual knows where to enter date, amount, and other information without having to search for the correct spot. These guides can be made at home or purchased from an optician or low-vision clinic.

Most optical aids are actually magnifiers of one type or another.

Some are used for reading and close work, and others are used for distance.

By far the simplest and most common optical aid employed for minimal to moderate visual loss is an augmented reading lens in one's spectacles. Instead of the usual strength reading glass or bifocal appropriate to the patient's age, a stronger convex lens addition is utilized. A reader wearing this stronger glass must hold the material at closer range for a clear focus. If moved further away where it is often more convenient to read, the material will be out of the focus of the glass (figure 12–4). Nonetheless, the closer focus provided by this stronger spectacle allows for the significant magnification that comes from bringing objects closer to the eye. Most people with reduced vision are willing to restrict the range of their vision in order to be able to see better at a closer distance.

But some may not benefit from stronger reading glasses because of occupational restrictions or vision that is too poor. Many individuals who could do well with augmented reading glasses simply do not like the restricted and close range. Hand-held magnifiers are a popular answer to this problem. A compromise is to supplement weaker bifocals or reading glasses with a hand-held magnifier for very small print or fine detail. The hand-held magnifiers come in a variety of shapes and powers that can be suited to the patient's needs and comfort.

Old people often have difficulty holding hand magnifiers at just the right distance from the page for long periods of time. Stand magnifiers are lenses set in a stand that is placed on the reading material and moved along. It is always at the correct focal distance from the page and is not affected by a tremor or fatigue. Some stand magnifiers and hand-held magnifiers come with built-in light sources as well. Adjustable reading and writing stands are available for those people who find it difficult to use magnifiers on a horizontal table.

Other handy devices are the neck magnifier, which is hung around the neck and supported against the chest; and the telemicroscope which is clipped or secured onto the front of a spectacle lens. The latter gives a higher degree of magnification than would otherwise be possible at this working distance.

As illustrated in figure 9–10, a problem that must be faced with

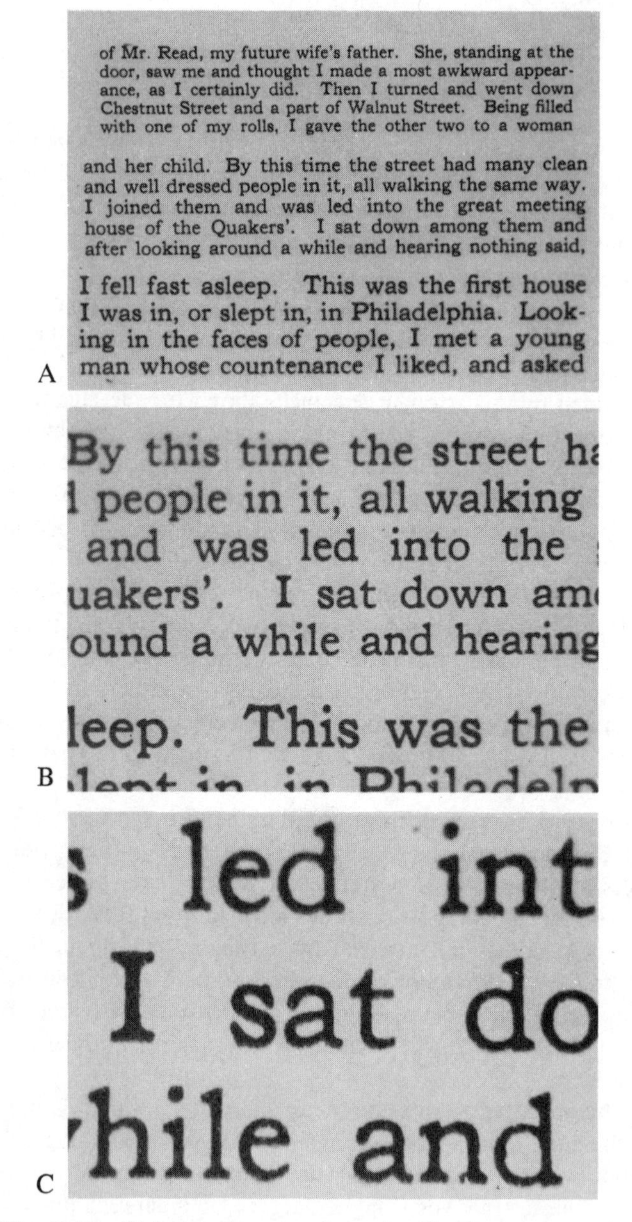

of Mr. Read, my future wife's father. She, standing at the door, saw me and thought I made a most awkward appearance, as I certainly did. Then I turned and went down Chestnut Street and a part of Walnut Street. Being filled with one of my rolls, I gave the other two to a woman

and her child. By this time the street had many clean and well dressed people in it, all walking the same way. I joined them and was led into the great meeting house of the Quakers'. I sat down among them and after looking around a while and hearing nothing said,

I fell fast asleep. This was the first house I was in, or slept in, in Philadelphia. Looking in the faces of people, I met a young man whose countenance I liked, and asked

A

By this time the street h
l people in it, all walking
and was led into the
uakers'. I sat down am
ound a while and hearing

leep. This was the

B

led int
I sat do
hile and

C

Fig. 9-10. Photographs A through C demonstrate increasing magnification resulting in a diminished field of vision.

optical aids for close work is that the visual field decreases with increased magnification.

Relatively new and still fairly expensive is the closed-circuit television apparatus adapted for low-vision work. Reading material is magnified into a closed-circuit television screen that can adjust brightness and contrast. Both black on white and white on black can be projected. In fact, some low-vision patients find white print on a black background easier to see.

These devices, some of which are pictured in figure 9–11, are

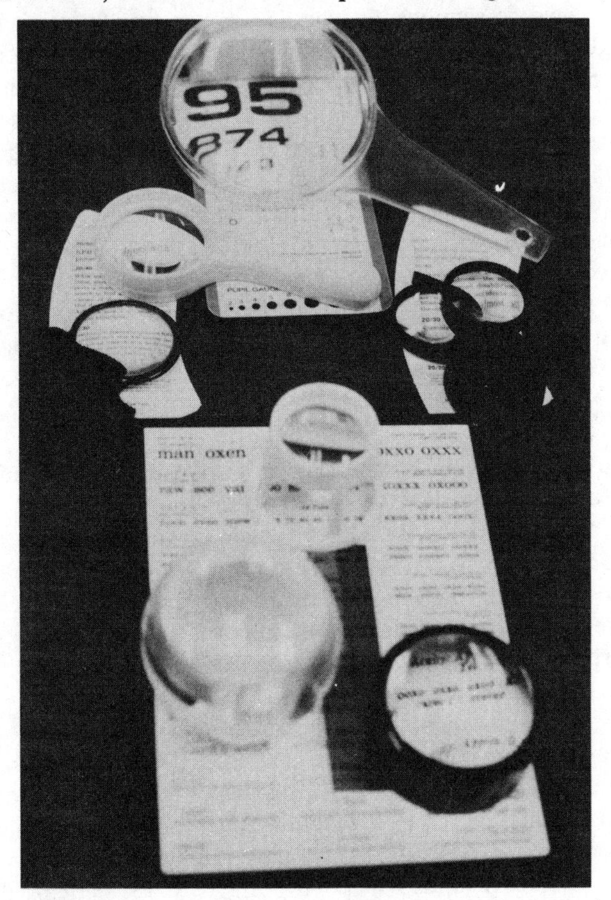

Fig. 9-11. Some of the more common visual aids used for reading and close work.

the most common aids prescribed. Many other optical aids not specifically designed for low-vision patients have been useful to them nonetheless. For example, a common device used in industry and in many medical offices is a large magnifying glass surrounded by a circular fluorescent light and mounted on a gooseneck stand.

Unfortunately, optical aids for distance vision afford less variety and flexibility. Small telescopes are the only devices that will magnify distant objects and allow a low-vision patient to see otherwise imperceptible detail. Because a telescope necessarily restricts the visual field so significantly, the devices are usually carried in the pocket and taken out for "spotting," such as checking the destination of a bus or reading a street sign. One cannot get around easily while looking only through a clipped-on telescope because of the loss of peripheral vision, but one can use the device to great advantage in a theater or in school for board work. Some telescopic lens systems have been miniaturized and adapted so that they can be incorporated directly into a spectacle lens. They are usually placed in the upper portion of the lens so that the wearer can tilt his head down slightly and glance through the telescope for a detailed view, then revert to the regular lens in order to maintain the full visual field. Figure 9–12 shows these telescopic aids.

Driving and Low Vision

Loss of a driving license can be a hardship for an individual whose occupation depends on it, or for one who lives in a rural area and depends on an auto for daily errands. It is not easy to decide when a visual disability is severe enough to disqualify an individual from driving. There are no good, widely accepted studies of the correlation between visual function and driving record. A limited study in Massachusetts showed that people who used telescopes to qualify for a license had no more, actually fewer, accidents than the average driver. The telescopes are incorporated into the upper part of a spectacle lens and used only for momentary spotting of signs, signals, or objects in the roadway, much as one glances in a rearview mirror. Looking through the telescope alone has the disadvantage, though, of restricting the driver's visual field, as illustrated in figure 9–13. Perhaps the surprisingly low accident rate reflects the practice and caution that are

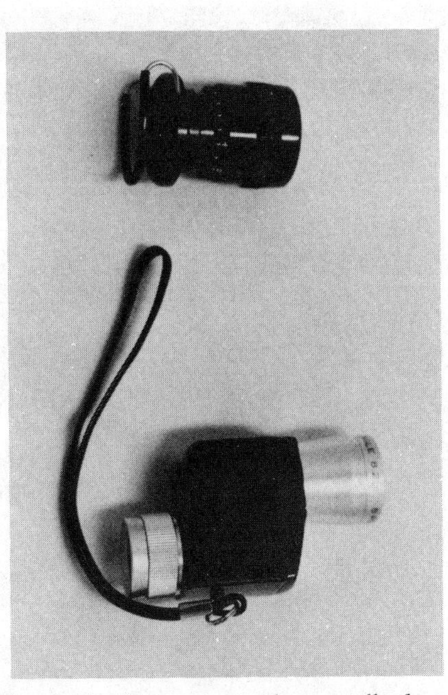

Fig. 9-12. Two commonly prescribed pocket telescopes used to improve distance acuity.

required of telescope users. In any event, the study was unfortunately limited and cannot be taken as proof that such a system is safe. North Carolina has a far more lenient visual requirement for licensure than most other states yet has no more accidents. But again, no real conclusions can be drawn. Driving conditions and individuals vary so greatly that generalizations or arbitrary criteria are usually inadequate. A congenitally affected individual with a stable level of low vision will be well adapted and far more capable than an elderly person coping with a sudden, even minimal, loss of vision. A person with good vision and poor judgment is far more dangerous than a very cautious but disabled one.

With no national guidelines, each of our fifty states has its own licensing criteria. In most states the minimum visual requirement is 20/40 in the better eye, but a North Carolina license may be obtained with 20/70 or less. Some states test color vision, while others do not even test visual field. Most states set more rigid

Fig. 9-13. A demonstration of the hazzard of restricted visual field through a telescopic device. The cars immediately to the left in photo B are not seen in the magnified view through a telescopic aid A.

criteria for a commercial license. Limited licenses to drive only during daylight, or only under 45 m.p.h., for example, are available in some states to individuals who cannot get an unrestricted license but still meet certain vision requirements. Presently, twenty-three states allow licensing for drivers who need telescopic lenses to pass the visual acuity test. The lenses should be prescribed by a competent eye doctor who will expect a trial period and at least a revisit to ensure that the patient understands the device and has acquired the reflexes necessary to use it properly.

Eye doctors are familiar with the licensing regulations in their states and can advise a patient of the requirements. While many criteria are presently arbitrary, it serves no one's interest to violate these well-intentioned laws until significant study and experience modify them. The resources listed in chapter 17 should be of interest to those who use or might benefit from optical aids.

10
Diabetes and the Eye

Diabetes mellitus is on the increase in developed countries due to recent progress in medical control and management of this disease. As a result, we are encountering more of the long-term complications of the disorder, one of which is the eye disease associated with it. Diabetic retinopathy now ranks as the leading cause of blindness in the United States. Cataracts and glaucoma are more common, but they are also more readily controlled.

Diabetes mellitus is a complex disorder of sugar metabolism. Glucose, the body's main energy source, can be utilized by the cells of the body only in the presence of the hormone insulin. Insulin is produced by specialized cells in the pancreas that regulate its release according to the energy needs of the body. Juvenile-onset diabetes, a disorder evident during childhood or adolescence, is characterized by a lack of insulin production. Treatment consists of insulin injections and carefully regulated diet and activity. Obesity may precipitate adult-onset diabetes, characterized by insufficient insulin. Oral drugs that step up insulin release from the pancreas have been used to treat the disorder, but their long-term complications have caused many patients to revert to insulin or more serious diet control.

But diabetes is a disease of more than just blood sugar and insulin production. Long-term complications may ensue even if the sugar level is carefully controlled with diet and insulin. Certain problems, particularly those associated with dysfunction of peripheral nerves, are related to the control. But the vascular complications, (premature arteriosclerosis, the vascular diseases of the kidneys and the eyes), appear to be more closely related to the duration of the diabetes than to its control.

Refractive Changes

The initial symptoms of diabetes—excessive thirst and urination, weight loss, and fatigue—may not be that prominent, particularly in adult-onset diabetes. Fluctuating vision and rapidly changing refractive errors sometimes appear as early symptoms of this disease. Glasses fitted recently may be inaccurate shortly thereafter, and changes may continue. The refractive errors result from changes in the lens of the eye caused by abnormal blood sugar levels. Unless one cannot do without them, it is not wise to prescribe spectacles until the diabetes is under treatment and relatively stable. The same principle is true for known diabetics who are temporarily out of control and are experiencing changes in refraction during this time.

Cataracts

Although diabetic cataracts are a recognized entity, they are rare. They occur in juvenile-onset diabetics between the teen-age years and the thirties. Overall, diabetics seem to develop the same senile cataracts common to the general population, though they develop a few years earlier than the average. With no significant retinopathy and with the diabetes under control, the prognosis for successful management of cataracts in diabetics is excellent. The same principles discussed in chapter 6 on cataracts in general apply to diabetics as well.

Cranial Nerve Palsies

The major nerves of the head and face are called the cranial nerves. There are twelve of these nerves, and three of them supply the muscles of the eye. Occlusion of a tiny blood vessel that supplies one of these nerves is a relatively common *microvascular* complication of diabetes. Double vision results from impaired movement of the eye muscles supplied by the involved nerve.

These nerve palsies are more common in the older adult-onset diabetics and are occasionally the symptom that leads to the discovery of diabetes. Spontaneous recovery usually occurs over a period of months.

Diabetic Retinopathy

The real threat to vision faced by diabetics is the associated retinal vascular disease. While an occasional rare patient will present first with this complication of diabetes, it most commonly develops many years after the diagnosis has been made. There is some controversy over what role the rigid control of blood sugar has in preventing problems like diabetic retinopathy and renal disease. Suffice it to say, no hard evidence at present supports the view that absolutely rigid control will avert such problems, though of course gross mismanagement and disregard for diet and therapy will cause many problems and exacerbate all abnormalities. At present, the most consistent correlation with the onset of retinopathy is the duration of the diabetes.

The exact cause for the blood vessel abnormalities in the retina is not known, but it is definitely at a cellular level. Certain supportive cells in the tiny capillaries degenerate and die, often leaving weakened outpouchings known as *microaneurysms.* Some of these capillaries block off altogether, leaving areas of the retina with an impaired blood supply; some leak fluid and lipids (fats); and still others burst and cause hemorrhages. The most serious aspect of retinopathy, however, may be a misdirected attempt of the retina to alleviate diminished blood supply. In a process referred to as *neovascularization,* fragile abnormal new vessels grow out of the retina or optic nerve head and into the central vitreous cavity. They are sometimes surrounded by fibrous tissue, and they often bleed, causing the eye to fill with blood in extreme instances. When neovascularization is present, the retinopathy is termed *proliferative*; otherwise it is known as *background.* The management and prognosis are quite different.

Background Retinopathy

Diabetic retinopathy is a disease of the retinal vessels. In background retinopathy one finds weakened and abnormal small vessels resulting in small retinal hemorrhages, microaneurysms, lipid deposits, excess fluid in the retina, and occluded capillaries (the smallest blood vessels connecting the arteries and veins). Figure 10–1 illustrates these findings in a typical case. A fluorescein

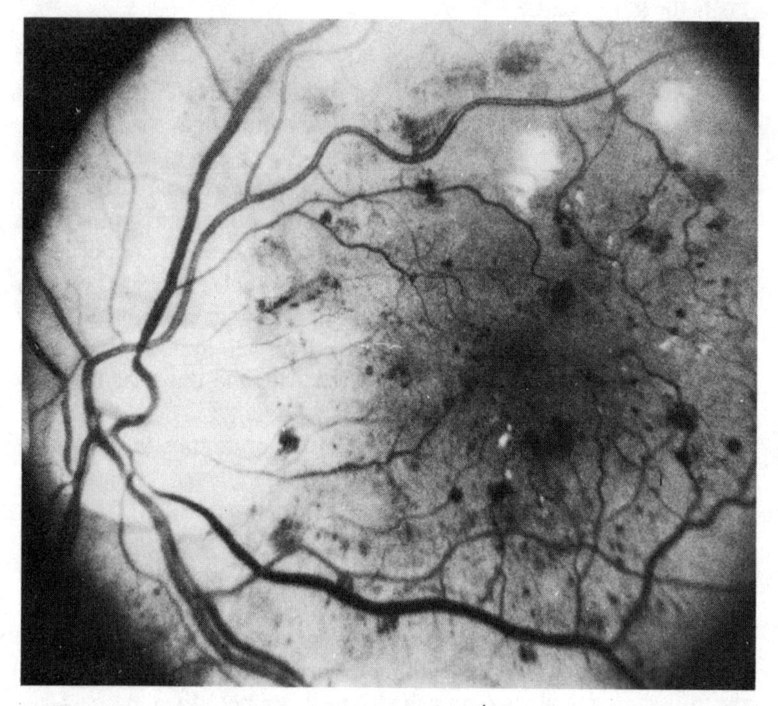

Fig. 10-1. Background diabetic retinopathy exhibits small hemorrhages and lipid accumulations in the retina, along with abnormalities of the small vessels such as microaneurysms, tiny beadlike weakened pouches from capillaries.

angiogram allows identification of the abnormal vessels. In this technique, fluorescein dye is injected into an arm or hand vein and a series of photographs taken (or the fundus of the eye observed with an ophthalmoscope) through a special cobalt-blue filter. This blue light causes the dye to fluoresce so that it can be traced as it moves through abnormal vessels or comes out of leakage sites. Figure 10–2 is a fluorescein angiogram of background retinopathy.

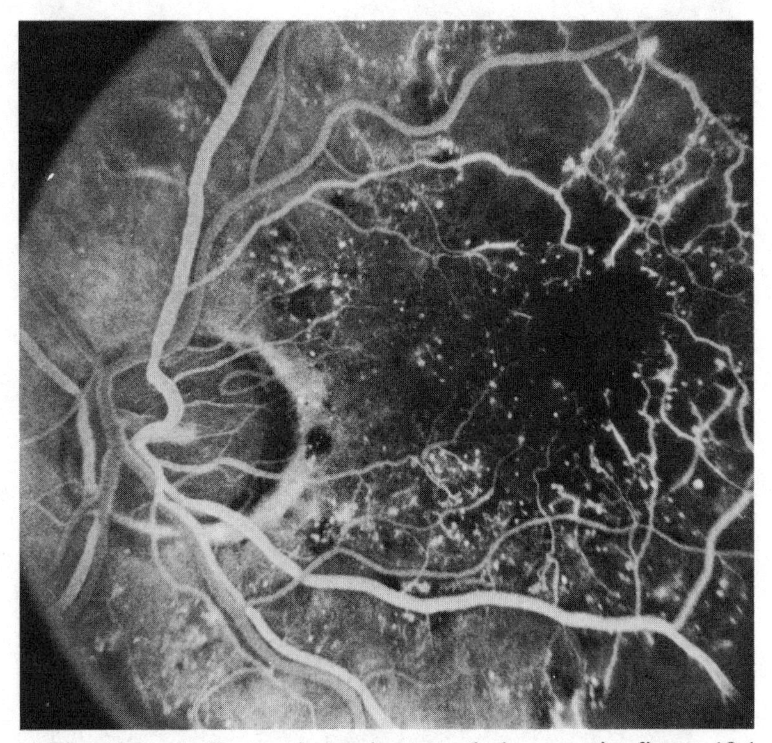

Fig. 10-2. A fluorescein angiogram of the case in figure 10-1 demonstrates vascular abnormalities and leakage sites common in diabetic retinopathy.

Fortunately, most cases of background diabetic retinopathy, despite their appearance to the physician, cause no symptomatic loss of vision. Only if a hemorrhage erupts or excess fluid accumulates in the macula will visual loss ensue. The macula is the pinpoint site of the retina responsible for keen central vision. *Macular edema,* as this fluid accumulation at that site is called, is actually the most common cause of visual loss in diabetics. Fortunately, it is usually a minimal or moderate loss which can be treated successfully with the laser in the majority of cases.

Figures 10–3 and 10–4 illustrate the ophthalmoscopic picture and fluorescein findings of a typical case of background diabetic retinopathy with macular edema. If such edema is allowed to per-

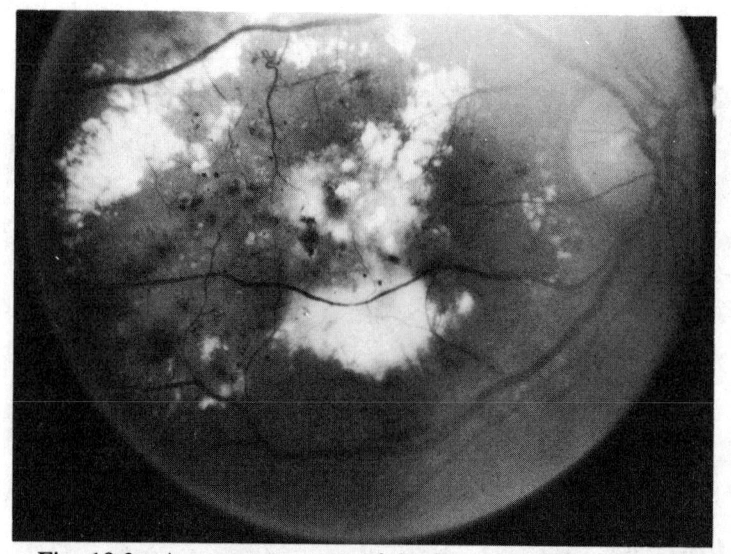

Fig. 10-3. An extreme case of background diabetic retinopathy with macular edema, an abnormal leakage of fluid that reduces vision.

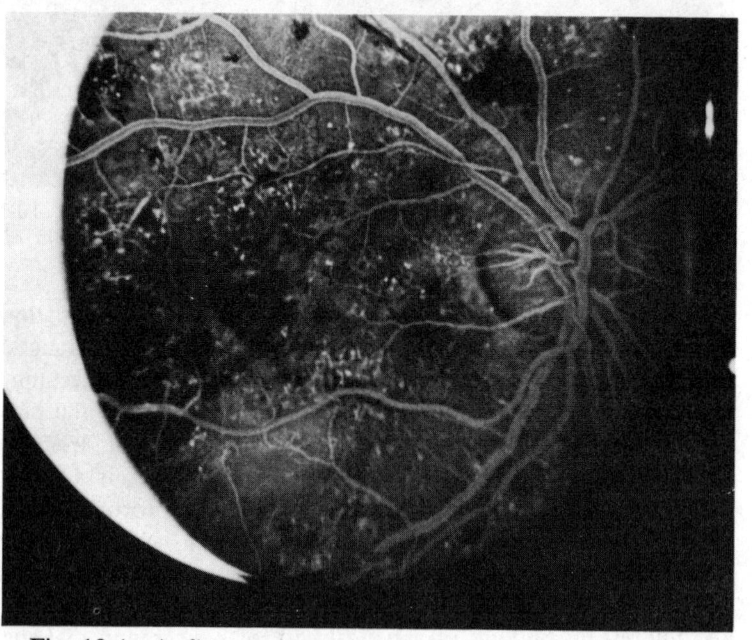

Fig. 10-4. A fluorescein angiogram of the case in figure 10-3 demonstrating early dye leakage from abnormal vessels.

sist, it may lead to heavy lipid deposits and permanent deterioration of vision. Occasionally, it will clear spontaneously. The disease must be monitored over the course of time, and the potential for improved vision with treatment must be considered.

Treatment

Most ophthalmologists now feel that treatment of background retinopathy is indicated only in the presence of diminished vision from macular edema, which on evaluation by fluorescein angiography shows a potential for return of vision. If the macular capillaries are occluded rather than just leaky, little or no return of vision can be expected. The laser treatment consists of placing discrete minimal burns at leakage sites to seal them off. The fluid that is already in the macula will then be absorbed, and with no continuing leakage vision will improve or at least remain stable.

Several oral medications that reduce the blood lipid levels have been used to reduce or eliminate the lipid deposits in the retina. While reduction in retinal lipid has been observed in some cases, there is no convincing evidence of significant improvement in vision. Most ophthalmologists evaluate background retinopathy with fluorescein angiography and consider laser therapy for the condition only when vision is diminished and macular edema is present.

Proliferative Retinopathy

Fortunately present in only a small percentage of diabetics, proliferative retinopathy is the most serious ocular complication of diabetes and can lead to blindness. In this form of retinopathy, occluded capillaries diminish the blood supply to patches of the retina. Deprived of its normal nourishment and ability to eliminate its waste products of metabolism, the retina somehow stimulates abnormal new blood vessels to grow. These fine new vessels break out of the retina and grow into the vitreous cavity. They are very fragile and will frequently bleed, creating a vitreous hemorrhage. This blood in the center of the eye may appear only as fine floaters or a small inkblot if limited, but it may also fill the vitreous cavity and block out all vision except bright light. The abnormal fibrous scar tissue that grows in the center of the eye after recurrent hemorrhages may actually pull the retina off into a traction detachment. Vision, then, can be lost not only from vitreous hemorrhage but also from scar tissue that blocks or detaches the retina. In its most advanced state, a total, scarred retinal detachment may result. Vision will be totally gone at this stage and the individual is blind. Figure 10–5 is a fundus photograph of neovascularization arising from the optic nerve head in a patient with long-standing diabetes. Figure 10–6 portrays the rapid leakage of the abnormal blood vessels.

Treatment

The treatment of proliferative retinopathy depends on the state of the eye at the time the disorder is recognized. Neovascularization is treated with laser or xenon light photocoagulation. Photocoagulation is the term for the use of laser or xenon (white) light to produce small, discrete burns in the retina. An extensive national study has shown laser therapy to be helpful in diminishing neovascularization and in preserving vision longer than would otherwise be expected given the natural course of the disease. During the three-year period of follow-up to date in the study, the incidence of severe visual loss was reduced by about 60 percent with treatment.

Laser therapy consists of scattering numerous minute burns over the periphery. Sometimes the abnormal vessels themselves are burned in an attempt to occlude them. Direct applications of laser to the abnormal vessels is somewhat more hazardous as it may precipitate a hemorrhage before occluding the vessel. The exact reason that neovascularization disappears after laser treatment to other areas is still a matter of conjecture. Perhaps it redistributes the blood flow or in some way alters the borderline metabolism of the retina so as to eliminate or reduce the production of chemical factors that stimulate the abnormal blood vessels.

Vitreous hemorrhages arising from neovascularization are usually small and spontaneously absorbed. Photocoagulation treatment cannot be carried out through the hemorrhage, as the light will be scattered by it. A new operation is available for the rare, extensive, or recurrent vitreous hemorrhages that do not clear. It involves inserting into the vitreous cavity a special aspiration, in-

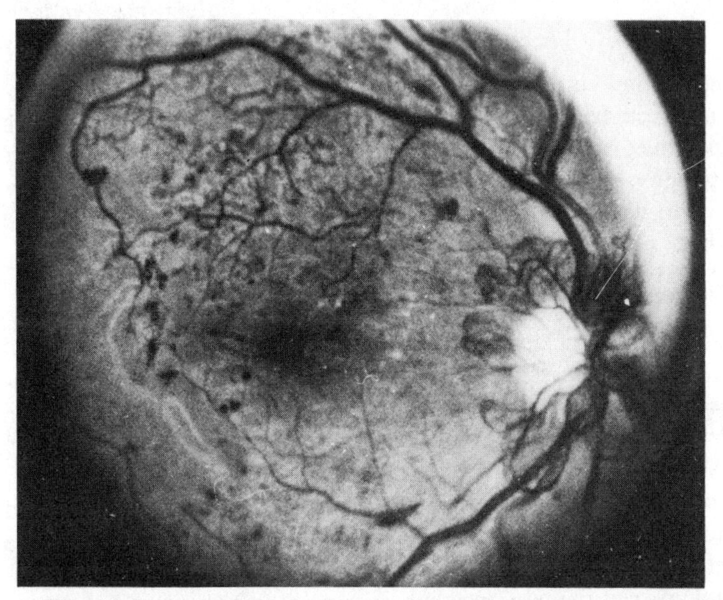

Fig. 10-5. Neovascularization, or abnormal new vessels, arising from the optic nerve head in a case of proliferative diabetic retinopathy.

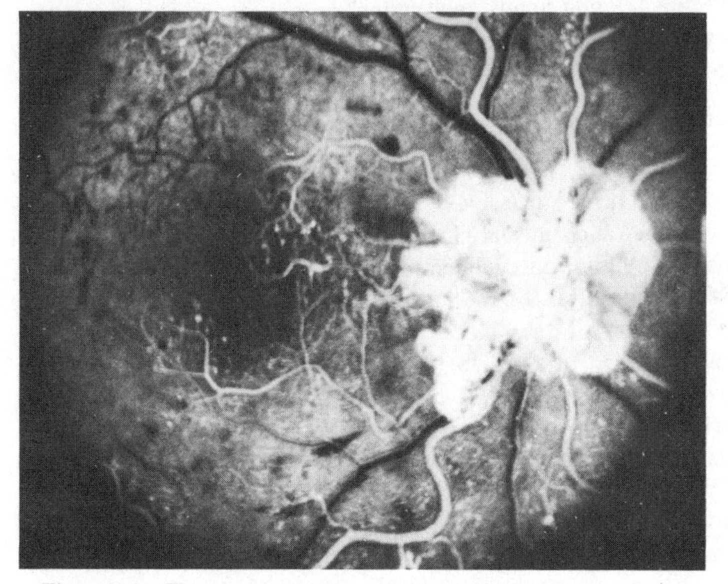

Fig. 10-6. Fluorescein angiogram of the case in figure 10-5 demonstrating rapid leakage from the abnormal new vessels.

fusion, and cutting probe through which the vitreous and hemorrhage can be removed. Because many hemorrhages will absorb spontaneously over a year or more, and because this vitrectomy operation does involve some risk, a nationwide study is now underway to determine which hemorrhages should be operated on and when.

Traction Detachments

The neovascularization that develops in proliferative retinopathy may be associated with fibrous tissue much like a scar. Repeated hemorrhages also result in scarring and fibrous tissue development in the vitreous. Fibrous tissue has a natural tendency to contract. If it adheres to the retina, it will pull the retina off. This sort of retinal detachment is called a *traction detachment* and is the most prevalent cause of permanent blindness from diabetic retinopathy. Among the most promising new advances in ophthalmology over the past decade have been the vitrectomy instruments that allow doctors to operate on these detachments by cutting the vitreous traction bands and reattaching the retina. Thanks to these new instruments, traction detachments no longer carry the nearly hopeless prognosis they did only several years ago.

Summary

In summary, diabetes is a disease involving more than just blood and urine sugar levels. It has many and varied effects on the eye. Rapid and unusual changes in refractive state may be due to undiagnosed or unregulated diabetes, while episodes of double vision or paralysis of the muscles that move the eye may be due to occlusion of tiny blood vessels that supply nerves to these muscles. Most of all, diabetics must be aware of the potential for developing retinopathy, a consequence of diseased retinal blood vessels. It usually presents no warning symptoms such as loss of vision, so periodic evaluation is prudent in order to disclose early —prior to visual loss—any potentially threatening retinopathy. Most ophthalmologists thus advise diabetics to have yearly eye examinations, even if asymptomatic.

11
Systemic Diseases and Medications with Ocular Side Effects

The eye is the only site in the body where blood vessels, nervous tissue (a direct outgrowth of the brain), and both voluntary and involuntary muscle can be easily evaluated by the physician. Because the eye is an integral organ of the body, a disease process that affects any tissue of the body also found in the visual system may affect the eye as well. Visual or ocular symptoms may be the first, sometimes the only, sign of a disease process affecting the body as a whole. Or a widespread process may have significant ocular effects that can be overlooked because of more obvious symptomatology or disease elsewhere. In addition, the eye is susceptible to side effects from many medications given for seemingly unrelated problems. This chapter will discuss several of the more important and more interesting disease processes that involve the eye as well as other parts of the body. It will also mention some of the more commonly used medications that can have insidious effects on vision and should thus be monitored with eye exams while therapy is underway. The number of diseases and rare syndromes with visual or ocular components is enormous and easily the subject of a large volume in and of itself; therefore no attempt to be comprehensive has been made. The number of drugs and therapies found to have visual side effects is equally staggering; a compilation is only as up-to-date as the latest drug included. Through his professional training and literature, your ophthalmologist is aware of the significance of any medications taken and of concurrent diseases or symptoms.

Nutritional Diseases

Malnutrition renders the body more prone to illness, and the eye is not exempt. Fortunately, in our developed and affluent society only very specialized circumstances and diseases lead to malnutrition. Protein and vitamin deficiencies still take their toll on the general population in less-developed countries, where it is not uncommon to see blinding eye problems that could have been prevented by ample nutrition. Chief among these diseases is *keratomalacia perforans,* a devastating thinning and melting of the cornea, the transparent front window, or crystal, of the eye. Measles, a disease that frequently attacks previously unexposed tribes or villages as an epidemic, causing a mild inflammation of the external eye, is often the final insult to an already weakened cornea. Infants and small children then develop corneal perforations, which can lead to loss of the eyes.

This problem is caused by vitamin A deficiency. In its less severe form, vitamin-A deficiency may lead to a drying of the external eye and roughening of its normally smooth, moist, mucus membrane covering, the conjunctiva. Night blindness and even death, in severe untreated cases, can result. In rice-eating countries where milk is inadequate and green vegetables are not incorporated into a child's diet, this preventable disease continues to ravage its victims. It is estimated that more than one percent of the preschool children in areas of Asia are affected.

The B-vitamin deficiencies cause *nutritional amblyopia,* a loss of vision due to degeneration of the optic nerve. The condition was prevalent in the prison camps of World War II, where pellagra and beriberi, diseases associated with B-vitamin deficiencies, were quite common. Loss of the central visual field and decreased color discrimination are usual symptoms. These problems seldom occur today; when they do, they are traceable to extreme dietary fads, intestinal ailments that affect absorption of the vitamins along with other nutrients, or other rare diseases which affect the body's utilization of them. The amblyopia exhibited by chronic alcoholics and heavy smokers is thought to be related to B-vitamin deficiency or its aberrant metabolism.

Vascular Diseases

In many cases, generalized diseases of the blood vessels of the body first become evident in the eye, since any problems relating to the tiny areas served by so few small vessels can significantly affect vision and perception. Thus, a tiny vascular occlusion or hemorrhage near the macula, the pinpoint of the retina responsible for keen central vision, will be immediately apparent to the patient, who will notice a loss of central vision in that eye. The retinal vessels, unlike any internal vessels elsewhere in the body, can be directly observed by the doctor. They are thus the best indicators of the status of the vessels of the body affected by a generalized disease process. The vascular system of the body can be divided into three parts. The arterial portion of the system consists of the vessels that bring fresh blood from the lungs and heart to the tissues. The capillaries are the very fine, microscopic channels fed by the arterial system. The capillary network spreads out through the tissue and is constructed so that gases, nutrients, and waste products of metabolism pass freely across its walls to interact with the tissue cells around it. The venous system is the return system of increasingly large collector vessels into which the capillaries empty so that the blood may return to the heart and lungs.

Some diseases of the vascular system affect preferentially the arteries, veins, or capillaries alone; others affect all components either directly or indirectly. The implications for the eye are quite different, depending upon the particular disease process.

Arteriosclerosis (Hardening of the Arteries)

A sign of arteriosclerosis is a broadening of the light reflex the examining doctor sees when viewing the retina's arterioles (small arteries). As the hardened arteries cross the tiny veins of the retina in the limited space available, they are surrounded by a common sheath. At these crossings the artery often slightly compresses the underlying vein. In extreme instances this compression of the vein leads to a cessation of blood flow in it and a venous occlusion results. The occlusion may occur in a branch vein in the retina or in the central retinal vein as the vein exits the eye in the optic nerve. Widespread retinal hemorrhages occur, along with excess fluid leakage into the retina from the backed-up veins and capillaries as shown in figure 11–1. The hemorrhages and excess fluid,

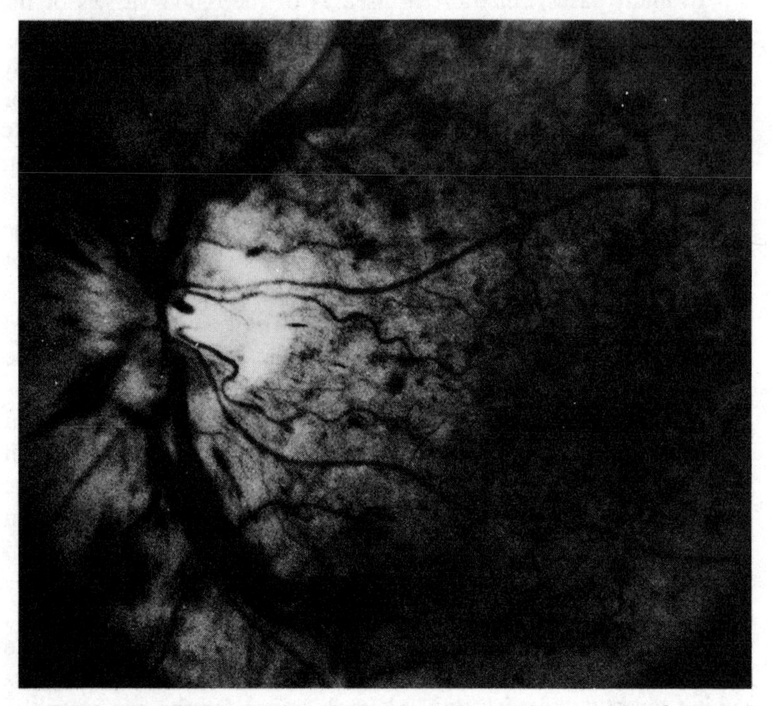

Fig. 11-1. Widespread retinal hemorrhages in a case of central retinal vein occlusion.

called retinal edema, will impair vision. If the condition does not clear spontaneously, the persistent edema may lead to tiny cystic cavities and degeneration in the retina, which can cause permanent visual loss.

With or without a venous occlusion, small hemorrhages and tiny arterial vascular occlusions can result from arteriosclerosis. The arterial occlusion may be of the main, central retinal artery, in which case all vision in the affected eye will be lost instantaneously. Or it may occur at the microscopic levels of the smallest arterioles, in which case the area of retina deprived of blood flow will be very small and unnoticeable unless it is directly in the central macular area that provides keen vision.

Vascular occlusions are a by-product not only of arteriosclerosis, but also of high blood pressure, diabetes, various inflammatory disorders of the blood vessels, and certain diseases that increase the viscosity, or thickness, of the blood. Because all of these disorders result in blockage of blood flow, we will discuss the significance of the occlusive disease that may result from any of them.

Central and Branch Retinal Vein Occlusion

Arteriosclerosis is the condition most commonly associated with retinal vein occlusions, but hypertension, inflammations of the blood vessels, abnormally high pressure in the eye from glaucoma, severe pressure on the vein after it exits the eye from an orbital tumor or hemorrhage, reduction of flow in the veins from leukemia or other blood disorders, and other processes that impair blood flow can all cause a venous occlusion. Ophthalmologists investigate all of these possible problems when they diagnose a retinal vein occlusion.

Many of the occlusions are only partial, and clear spontaneously; others persist or do not clear enough to relieve the retinal swelling (edema) that results. If there is evidence of further deterioration of vision or irreversible degenerative changes in the retina, therapy must be initiated.

A fluorescein angiogram is usually performed: a fluorescent dye injected into a hand or arm vein is then photographed and/or observed as it passes through the vessels of the eye. The site of blockage and extent of leakage can then be evaluated. Retinal photocoagulation, small burns made by a laser or a special xenon light source, is utilized to seal leakage sites and/or diminish the blood inflow. It is a painless procedure in which a light beam is aimed into the eye while the patient either lies down or is seated at the instrument. It is no more uncomfortable than a retinal examination with an ophthalmoscope. Local anesthesia is rarely needed.

In addition to retinal edema, another complication that can arise after a venous occlusion is neovascularization (figure 11–2). Fragile new blood vessels grow out of the retina and into the

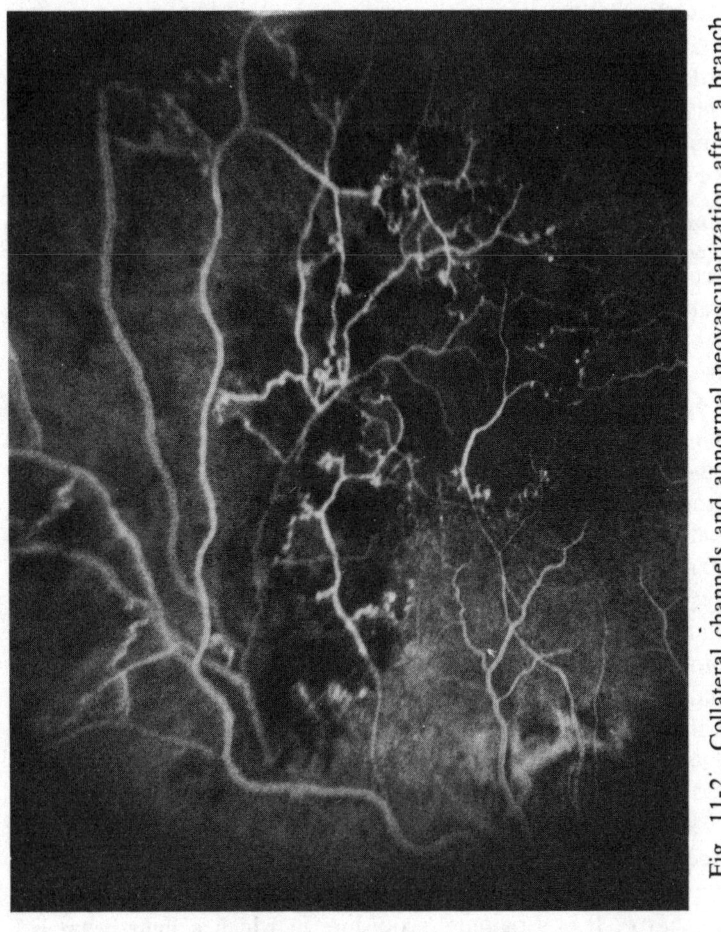

Fig. 11-2: Collateral channels and abnormal neovascularization after a branch vascular occlusion (fluorescein angiogram).

vitreous cavity, apparently after being stimulated in some way by the altered metabolism or blood flow. These new vessels frequently bleed, causing vitreous hemorrhage, which can severely diminish vision if blood fills the posterior chamber of the eye. New, abnormal blood vessels may also form on the iris in the front of the eye; when they form at the base of the iris in the filtration angle where the aqueous fluid exits, they can impede aqueous outflow and provoke severe glaucoma. This cause of high pressure in the eye, called *neovascular glaucoma,* is very difficult to control. Neovascularization is treated with photocoagulation. Glaucoma is treated medically or, if necessary, surgically. All patients who have had a venous occlusion must be monitored and evaluated for these potential complications, which may develop up to six months or more after the original vascular problem occurred.

Central and Branch Retinal Artery Occlusions

The symptom of a central retinal artery occlusion is sudden overall loss of vision in the affected eye. A branch arterial occlusion may result in obvious central visual loss if the macula is involved, but usually only one quadrant of the vision in one eye is blocked out. Retinal arterial occlusions are also most commonly associated with arteriosclerosis, in which the center of the arteriole (small artery) becomes increasingly narrowed as lipid deposits and degenerative material build up inside the vessel walls. Finally, the blood flow becomes so disturbed in these narrowed areas that it clots, forming a *thrombosis.*

Temporal Arteritis. A rarer but more treatable cause of central retinal artery occlusion is a peculiar inflammation of the arteries known as *temporal arteritis* or *giant cell arteritis.* It commonly involes the artery leading to the temple, which becomes inflamed and is characterized on microscopic examination by special "giant" cells. Typical symptoms are severe pounding headaches in the temple area as well as bulging and tenderness of the artery. Pain may develop in the jaw muscles after chewing. Victims may suffer low-grade fevers, diffuse muscle aches, malaise, and weight loss. Sudden, irreversible loss of vision from central artery occlusion often accompanies this disease, since the arteries to the optic nerves also become inflamed. Blindness is one of the disease's worst complications. Fortunately, the disease can be countered effectively with steroid medications, but they must be started at once in order to prevent the second eye from going blind also. A simple and quick lab test is usually sufficient to determine whether or not an individual has this disease. If questions arise, a biopsy of the superficial temporal artery can be taken to confirm a diagnosis. The procedure is harmless.

Severe high blood pressure may also cause arterial occlusions in the eye. In such cases the arteries become so narrowed from spasm that blood flow diminishes and finally ceases. Even though blood flow is reduced, retinal swelling may occur because the retinal cells can no longer function properly without nutrients. Vision is lost for the swollen retinal cells cannot function. Because visual loss is often the first sign of severe hypertension, it may first be diagnosed by the ophthalmologist, who notes the characteristic appearance of the narrowed retinal vessels, retinal hemorrhages, swelling, and arterial occlusions.

Emboli are tiny fragments that get into the blood stream and travel along until they reach an arteriole too small to pass through. They lodge at this point in the arterial system and cause an occlusion and interruption of the blood flow beyond. Emboli come from several sources. An infection located anywhere in the body, particularly on a valve in the heart, may release tiny clumps of organisms and white blood cells into the bloodstream. A plaque from arteriosclerosis located on the inner surface of a major artery may fragment and drift along with the blood flow until it lodges in a smaller arteriole. Part of a remote blood clot, or fat from bone marrow in a broken bone, may also enter the blood stream as an embolus. The eye is the only site in the body where these emboli can be seen directly. Thus, a diagnosis of an infected heart or of a major arteriosclerotic plaque in the carotid artery may emerge from an examination of the eye.

The treatment of central retinal artery occlusion is intended to restore blood flow and oxygenation in the retina. If the artery becomes completely blocked, irreversible damage can take place in a matter of minutes. It is wise to know simple emergency maneuvers that can suffice until a physician is reached.

The blood vessels of the brain and eye respond to a build-up of carbon dioxide in the blood by dilating reflexly. A build-up of carbon dioxide and a concomitant decrease in the necessary oxygen indicate a failure of circulation or respiration. The arteries dilate in an attempt to compensate.

A central retinal artery occlusion may be responsible for a sudden painless overall loss of vision in one eye. Steps may be taken immediately to reestablish blood flow. Having the patient breathe in and out of a paper bag held over the nose and mouth will increase the carbon dioxide level in the blood since exhaled carbon dioxide will be reinhaled. Momentary, gentle pressure on the eye may dislodge an embolus or move it along, but the patient will need prompt medical attention. An ophthalmologist may give the patient a special mixture of high carbon dioxide, high oxygen gas to breathe. Medicines or injections will quickly lower the pressure within the eye and relax arterial spasms. A doctor will then determine whether the occlusion was caused by arteriosclerosis, a clot or an embolus, high blood pressure, temporal arteritis, or other inflammatory conditions.

Unfortunately, most patients with complete arterial occlusion seen by ophthalmologists have already suffered irreversible retinal damage. Studies have shown that therapy can be helpful if initiated within hours of the occlusion. After a day or more it is fruitless.

Diabetes is one of the most common systemic diseases seen by ophthalmologists due to its ocular and retinal vascular complications. Because of its great importance and increasing frequency, a separate chapter has been devoted to it.

Rheumatoid Arthritis

Rheumatoid arthritis is a deforming, painful, and crippling disease that causes swollen, red, painful joints. It also may provoke fever, anemia, ocular symptoms, and other problems. Newer treatment regimens are proving more effective in controlling the symptoms of rheumatoid disease, but neither cause nor cure has yet been found.

About 15 percent of the patients with rheumatoid arthritis suffer from dry eyes. This decrease in tear formation, often associated with a dry mouth due to diminished salivation, is called *Sjogren's syndrome.* Roughly half the people who consult an ophthalmologist for treatment of dry eyes and mouth are later shown to have rheumatoid arthritis or a related inflammatory disease of connective tissue, joints, and small blood vessels.

Dry eye feels to the patient as though a foreign body, such as a grain of sand, has lodged in the eye. The eyes may become inflamed at times, and symptoms are generally worse on dry, windy days. Chapter 4 discusses the treatment of dry eye. Simple artificial teardrops usually suffice.

Few individuals with rheumatoid arthritis develop a severe inflammation of the sclera, the thick white coat of the eye. This situation requires immediate attention since it may be due to an occluded vessel, in which case the sclera will thin and even perforate. Adjacent areas of cornea may become inflamed and melt. If medical therapy does not help, an operation to place a patch graft of sclera over the perforation may be necessary to save the eye.

Rheumatoid Spondylitis

Rheumatoid spondylitis is a rheumatoid disease. But it differs from rheumatoid arthritis in that it affects younger males between twenty and forty almost exclusively, as opposed to rheumatoid arthritis, which affects older individuals who are more likely to be female. It affects primarily the spine, and its eye complications are quite different. It is most commonly associated with an inflammation within the front section of the eye, a condition called *iridocyclitis.* The symptoms consist of redness, pain (especially on exposure to light), and blurred vision. Scarring of the pupil, glaucoma, cataract, and other problems may result. Any person known to have rheumatoid spondylitis, also called *ankylosing spondylitis* and "Marie Strumpell disease," should have an eye evaluation by an ophthalmologist. And any eye symptoms or irritation should be checked immediately.

Juvenile Rheumatoid Arthritis

This variant of rheumatoid arthritis occurs in children and has its own constellation of distinguishing features. It can present with tender, red, swollen joints, fever, skin rash, and other problems. Eye complications are its most insidious feature. Significant inflammation within the eye occurs frequently, but often without visible symptoms. The eye is usually white and appears normal. Cataract and other problems may develop, so it is imperative that a child who has been diagnosed as having juvenile rheumatoid arthritis have an ophthalmologic exam and periodic checkups. Between 5 percent and 15 percent of the children who initially develop this disease in multiple joints also develop eye complications. Oddly, for no apparent reason, about twice that percentage of children develop ocular problems if their disease initially involved one joint.

Thyroid Disease

The thyroid gland regulates body metabolism through its hormone, thyroxine. Production of this hormone by the thyroid gland is governed by the pituitary, and this gland, in turn, is influenced by the brain. Various feedback and regulatory mechanisms keep the body finely tuned. When one of these mechanisms goes awry, or the gland itself malfunctions, the thyroid may underproduce or overproduce its hormones. These conditions are called hypothyroidism and hyperthyroidism, respectively. In the former, inappropriately low amounts of hormone lead to weight gain, sluggishness, and other symptoms. The hyperthyroid individual (one who has an overactive thyroid) suffers weight loss, general nervousness, and irritability, among other symptoms.

Several eye complications of thyroid malfunction are difficult to relate to the control of the thyroid gland itself. Because of the complex interactions involved, intermediate hormones and substances may still be unregulated and may cause ancillary problems, even though therapy directed at the thyroid gland may have its obvious malfunction under control. Common eye problems that accompany thyroid disease are upper lid retraction, proptosis (bulging) of the eyes, diminished ability to converge the eyes, and extraocular muscle weakness and restriction of movement. The classic picture shown in medical texts is that of a young, thin female patient with a goiter (swollen thyroid gland) and bulging, staring eyes. This hyperthyroid goiter with *exophthalmos* (another name for proptosis, or forward bulging of the eyes) is called *Graves' disease*. The eyes bulge because there is an excessive deposition in the orbits of a substance commonly found in connective tissue spaces of the body. Often, when medication, surgery, or X-ray therapy reduces the activity of the thyroid gland, the eyes recede to their normal position. The same bulging of the eyes can occur with underactive thyroid glands; sometimes the proptosis affects one eye only. If severe enough to limit the ability of the eyelids to close over the eye, the condition can result in an irritating and dangerous exposure of the cornea. Infrequent blinking or an inability to close the eye completely will dry out the cornea and conjunctiva and render them susceptible to infection and foreign bodies. If frequent application of artificial tears by day and ointment and taping of the lids at bedtime do not suffice, an operation that partially closes the lids may be necessary. In extreme cases, large doses of anti-inflammatory steroid medication may be administered in an attempt to cause the eyes to recede. If this medication does not work, an operation that removes the outer (or any other) wall of the orbit may be performed in order to decompress it and allow the eye to fall back.

Occasionally, an inflammation of the muscles that move the eye is associated with thyroid malfunction. The inferior rectus, or lower, muscle is usually involved. When the muscle's action is limited, a malalignment of the eyes and double vision can result. If prism glasses do not help and there is no sign of spontaneous improvement, an operation can realign the eyes in the most important fields of vision. The muscle is recessed to loosen it and to allow the eye to assume a more normal primary position.

Myasthenia Gravis

Myasthenia gravis is a disorder that affects the transmission of nerve impulses to muscles. A chemical is released at the end of a nerve when an impulse is transmitted along it; this chemical then acts on special receptor sites of the muscle to initiate muscle contraction. The reason this transmission from nerve to muscle becomes faulty is not fully understood, but the condition results in weakness and, if severe enough, in an inability to move the muscles at all. The eyelid and eye muscles may be involved in this process; occasionally they are the first muscles to be visibly affected. A drooping eyelid or double vision from malaligned eyes incapable of a full range of motion brings the patient to the eye doctor. Typically, the weakness and double vision worsen later in the day. The doctor may perform a diagnostic test by injecting the patient with a medicine known to augment the action of the transmitter substance from nerve to muscle. If the weakness and symptoms improve, the test is considered "positive." The first line of therapy for this disease is the administration of an oral medicine, which has this same effect.

Diseases of the Brain and Nervous System

The eyes are a direct outgrowth of the brain, as are the nerves that regulate the eye muscles and lids and give sensation to the cornea, external eye, and lids. These *cranial nerves* are distinguished from *peripheral nerves,* which connect with the brain through the spinal cord. There are twelve cranial nerves, six of which are directly related to the function of the eyes. The visual pathways, the tracks of nerves that carry the visual perceptions of the retina to the brain, travel from the front of the brain to its most posterior tips, the *occipital lobes,* where we translate these impulses into vision. It is easy to understand why so many disorders of the brain and nervous system affect the eyes or visual system in some way.

Multiple sclerosis, for example, is a disease that results in an inflammation and patchy loss of the nerve's insulating sheath, called *myelin.* It commonly affects the optic nerve and causes a sudden loss of central vision in the affected eye, usually accompanied by pain when the eye is moved. Or it may affect the nerves that regulate the eye muscles, resulting in double vision. The initial attacks of this disease are usually transient, with more or less complete recovery of function. A recent long-term study of multiple sclerosis is far more optimistic in outlook than was previously justifiable. Some twenty-five years or more after the diagnosis was established, most patients were still able to get around alone and function well.

Other afflictions of the nerves can produce ocular signs and symptoms identical to those of multiple sclerosis. Optic neuritis, for example—the loss of vision due to inflammation of the nerve

—may be caused by viruses like influenza and mumps, by a small vascular occlusion to the nerve (a stroke of the optic nerve), by other rare inflammations of the nerves, by toxins such as methyl alcohol, and even by B vitamin deficiency. These problems can affect other nerves as well and result in impaired ocular motion, double vision, decreased sensation, inequality of the pupils, and blurred vision.

Tumors, aneurysms, abscesses, and other space-occupying disorders that impinge upon any of the nerves may also cause malfunction. If they involve the visual pathways in the brain, they will produce characteristic visual field defects, depending on their location. For example, a brain tumor (or occluded blood vessel or "stroke") involving the upper-left side of the visual pathway traveling to the upper-left occipital lobe disturbs nerve fibers that carry impulses originating in the upper-left visual quadrants of each eye. Because the eye projects the light that enters it upside down and backward on the retina, the field of vision lost is the lower right quadrant in each eye (figure 3–6B).

If a mass such as a tumor or abscess or a diffuse swelling of the brain from trauma or toxicity occurs, the pressure within the skull may become elevated. This elevated pressure is transmitted along the optic nerves within their outer coats, which are continuations of the coats of the brain. When pressure in the head is increased, the optic nerve begins to bulge forward into the eye. Engorgement of the optic nerve due to increased intracranial (within the skull) pressure is called *papilledema.* It may be the first evidence of a tumor or swelling and may be noticed incidentally during a routine eye exam. Doctors always look for it when evaluating headaches or suspicious neurological problems.

Blood Disorders

Various disorders of the blood, such as leukemia and diseases of the lymphatic system, produce hemorrhages and sometimes thickening of the blood, which causes reduction of flow. These diseases may be evident in the eye as retinal hemorrhages, distended veins, or actual retinal vein occlusions. Accumulations of tumor cells sometimes invade the orbit or tear gland. These symptoms are not usually the first manifestation of such diseases.

Hemoglobinopathies also affect the eyes, and ocular involvement in these disorders may be their most prominent symptom. Genetic differences in the structure of hemoglobin, the large protein that carries oxygen in the red blood cells, are presumably developed through natural selection. A variant of normal A-type hemoglobin developed and spread in areas of Africa because it rendered the red blood cells less susceptible to malaria, a serious disease common in those areas. But this variant also resulted in anemia and distortion of the red blood cell into a sickle-shaped cell that could occlude small vessels. This disorder is called *sickle-cell anemia.* Victims of this disease have two genes for sickle hemoglobin (SS), one from each parent. They are anemic and suffer from painful crises in which small blood vessels block off. Transfusions of normal blood are often necessary. Vascular occlusions may occur in the eye as well, but significant ocular problems are surprisingly rare. Individuals who have only one sickle cell gene are said to be carriers and are estimated to compose about 8 percent of the American black population. Under rare circumstances, such as severe dehydration, very low oxygen pressure (present at high altitudes and in nonpressurized aircraft), or ocular trauma and high pressure in the eye from secondary glaucoma, otherwise normal carriers may develop occlusions of the blood vessels due to sickled red blood cells. But ophthalmologists more commonly observe hemoglobin SC, another hemoglobin variant. This hemoglobin is a combination of sickle hemoglobin and another abnormal type that is labeled C. Perhaps because this abnormality causes slower, partial vascular occlusions than SS, and thus liberates some factor that stimulates new vessel growth, or perhaps due to influences that are not yet fully understood, it results in peripheral retinal neovascularization. In the anterior part of the retina, in response to occlusion of the narrower retinal arterioles in this area, fragile new abnormal blood vessels develop. These vessels tend to bleed easily, which results in recurrent vitreous hemorrhages. While patients with SC hemoglobinopathy may have vascular occlusions elsewhere in the body due to extenuating circumstances, the ocular problems are often the most prominent symptom. Photocoagulation is the most common treatment used to seal off these abnormal new vessels.

Skin Diseases

The cornea, conjunctiva, and lens all derive from the same primitive ectodermal cells that give rise to the skin in the developing embryo. Thus the eye is often involved when an individual succumbs to certain skin disorders. A discussion of the more common of these diseases follows.

Lid Tumors

In chapter 4 we discussed the eyelids and the types of infections they are susceptible to. The eyelids are also prone to skin tumors; one of the most common, the basal-cell tumor, is often found there. This tumor is an overgrowth of one of the cellular layers of the skin. Usually developing on the lower lid, it grows slowly and almost never seeds the blood stream or lymphatic system to initiate remote secondary tumors (*metastases*). This tumor is locally invasive and destructive, however, and should be treated as soon as it is discovered. It usually appears elevated with rounded, pearly edges and a small crater in the center. Occasionally, it ulcerates and bleeds. It is rarely more pigmented than the surrounding skin. Excess pigmentation suggests another, more malignant skin tumor called a melanoma. There is so much variation in these skin disorders that even experts are fooled sometimes by appearance alone. Only a biopsy can establish a diagnosis with certainty. Thus, any questionable skin mass or sore which fails to heal rapidly should be brought to a physician's attention. If it is small enough, an *excisional biopsy* can be performed under local anesthesia, and it can be totally removed. For larger lesions, an inconspicuous biopsy may be performed to indicate whether wide excision, irradiation, or other therapy is necessary for tumor. Simpler procedures may be sufficient for a benign mass.

Seborrheic Dermatitis

Seborrhea causes excessive oiliness of the skin. Dandruff is present and red, scaly patches develop in skin folds, behind the ears, alongside the nose, in the brows, and elsewhere, in extreme cases. This condition often affects the eyelids as well; they become red, swollen, and crusted. Secondarily, the conjunctiva and cornea become irritated and the eye also becomes red. Infection may supervene in the already inflamed location. General treatment, such as antidandruff shampoos and skin medication, seem to help clear up the lids. But anti-inflammatory steroid drops or ointments may be necessary also. If infection is present, appropriate antibiotics may be added. But these topical eye medicines should be used only for flare-ups. The disease cannot be cured by these measures, and it may be better controlled without eye medicines by simple lid hygiene measures. Cotton-tip applicators dipped in a sudsy mixture of baby shampoo and warm water can be used to wash the lids along the margins and at the base of the lashes. Care must be taken to pull the lower lid down and away from the eye while washing it. One must wash by feel rather than by sight. One should actually look up and away so as to move the cornea away and minimize the slight risk of rubbing the eye. Similarly, the upper lid should be pulled up while one looks down in order to wash it. Pressure should not be placed on the eye itself, but rather on the bone of the brow or cheek when pulling the lids away. This washing eliminates scales and oil in which bacteria proliferate, and often will even alleviate the inflammation (figure 4–2 and 4–3).

Psoriasis

Psoriasis is similar to seborrhea, but it is a dry, scaling inflammation. It, too, can affect the lids and lead to secondary infections, eye inflammation, and scarring. Lid hygiene measures as described above and topical medications for the eye may be needed to control the lid problem.

Rosacea

This common disease of middle-aged and older individuals is characterized by dilated blood vessels in the face, red, acne-like eruptions of prominent oil glands of the facial skin, and a coarse, porous thickening of the skin of the nose. The lids and conjunctiva are often chronically inflamed. In more serious cases, the cornea may be inflamed and scarred. Bacterial infection is quite common with this disorder, and oral as well as topical ocular antibiotics may be necessary. An ophthalmologic evaluation and periodic checkups are wise if the lid or eye is involved.

Atopic Dermatitis

Atopic dermatitis, also called eczema, is an allergic inflammation of the skin that may attack the eyelids in particular. This disorder is also infrequently associated with cataracts, retinal detachments, and corneal disorders. It is prudent to have an eye examination if significant lid dermatitis is present or if any ocular symptoms develop. Anti-inflammatory ointments provide some relief.

Sarcoidosis

Sarcoidosis is a disease that affects many areas of the body in a pattern similar to tuberculosis. But no microorganism has been identified in sarcoid and thus the cause is not known. It invades the lungs, the liver, the lymph nodes, the nervous system, the skin, and many other areas. It can cause severe inflammation in the eye, a condition known as *uveitis.* It is usually symptomatic, causing the eyes to become red, painful and extremely sensitive to light. The disease can also be less obvious and cause only mild blurring of vision or minimal redness. Disaster may be imminent, however, since dense inflammation within the eye can cause scarring and distortion of the pupil, glaucoma, cataract, vitreous clouding, and retinal damage. Any patient with known sarcoidosis should have a comprehensive eye examination to determine the extent of ocular involvement; any patient with a suspicious ocular symptom should see an ophthalmologist immediately. Treatment with topical medications and systemic anti-inflammatory agents usually arrests the ocular inflammation and prevents severe complications.

Systemic Medications with Ocular Side Effects

Most potent medications have myriad side effects and complications. Obviously, control of the disease process for which the drug is used must justify the risks assumed. We will discuss several common medications that require periodic eye examinations when used regularly.

Corticosteroids

The corticosteroids are derivatives of the hormones produced by the outer layers of the adrenal gland. These potent medications have revolutionized medicine by controlling many poorly understood diseases and problems. But they also have a host of side effects and complications. Prednisone and its derivatives are potent anti-inflammatory agents used to control destruction caused by inflammation. They are also used to relieve allergic problems such as asthma, and as part of the therapy for lymphomas and leukemias. The most frequent and significant eye complication is cataract formation. It is not a common complication, given the large number of individuals who take or have taken steroids, but it is worth watching for if large dosages or chronic low-dose therapy are employed. The cataract might not progress if it is discovered early. An alternative treatment can be employed or the dosage lowered. As with any medicine, some individuals are undoubtedly more susceptible to certain effects than others.

Chloroquine

Chloroquine is a quinine-like drug originally used to treat malaria. But lately it has been found to have potent anti-inflammatory effects in such conditions as rheumatoid arthritis and lupus. Chloroquine is absorbed and stored by the pigment cells of the body, including the important pigment epithelium layer in the retina. Thus, after long-term use of a substantial amount of this agent, retinal toxicity can result. Diminished vision, impaired color discrimination, and difficulty with night vision may develop. The more chloroquine one has taken, the more likely it is that toxicity will develop. Several years of continuous use usually pass before a significant cumulative amount of the medicine is reached. But chronic, continuous therapy for many years is not uncommon, and it is wise to have an initial baseline eye examination and then periodic checkups as advised to monitor any suspicious changes. Visual loss from toxicity is irreversible, but it will not progress if the use of chloroquine is discontinued. It is best to have the ophthalmologist discover the first subtle signs of toxicity.

Ethambutol, INH (Isonicotinic Acid Hydrazide)

These two drugs are used to treat tuberculosis. The therapy is usually necessary for a year or more, and at any time during the use of these medicines a deterioration of optic nerve function may develop. This side effect is quite rare and seems to be a function of individual susceptibility. It is wise to have a baseline pretreatment eye exam in order to determine what level of vision and what ocular conditions exist so that any later change can be recognized and understood. Patients should have periodic visual checkups and should check vision themselves each week. They should report any change immediately to the physician. Vision usually returns when the drug is discontinued.

Vitamins A and D

The old adage "if some is good, a lot is better" does not apply to these vitamins. Faddists or ill-advised individuals who take excessive amounts of these vitamins often develop significant side effects. No side effects have been reported by persons who take recommended dosages; they are reported by those who take grossly excessive amounts without benefit of a physician's advice.

One can actually develop brain swelling and increased pressure within the skull by taking too much vitamin A. Double vision due to impaired eye movements, retinal hemorrhages, and loss of lashes and brows may also occur.

Excessive vitamin D intake can cause abnormal calcium deposits throughout the body. These deposits may develop in the conjunctiva and cornea and cause a horizontal band of degeneration across the center of the surface of the eye. This *band keratopathy* produces loss of vision and extreme irritation.

Tranquilizers

Some of the common tranquilizers may cause such ocular side effects as increased pigmentation of conjunctiva, cornea, and lens, retinal and/or macular degeneration, and occasionally a temporarily decreased ability to accommodate and focus near objects. If long-term therapy at significant dose levels is contemplated, or a significant cumulative amount of these agents has been ingested over a long period of time, one should undergo an ophthalmologic evaluation.

Antidepressants

The most common ocular side effect of these medications is an impairment of accommodation, a decrease in the ability to focus near objects. This so-called complication is usually an actual pharmacologic effect of the drug and will disappear when the effect of the drug dissipates. If necessary, reading glasses or bifocals can be obtained or adjusted in order to compensate for this effect.

Antispasmodics

These agents are commonly used to reduce intestinal spasm and pain in such conditions as peptic ulcer, gastroenteritis, and colitis. They are atropine-like drugs which may cause diminished accommodation with inability to focus at close range. They may also cause dilation of the pupil and the attendant risk of acute angle-closure glaucoma in susceptible individuals. It is rare, however, to experience these systemic effects from an occasional dose that does not exceed the recommended amount.

A general comment about side effects from drugs is in order. Even water, if taken in excess or under certain extraordinary circumstances, can have deleterious effects. A listing of all the side effects of even common, relatively benign medications would sound terribly ominous. Decongestants used for colds or upper respiratory infections, for example, may induce glaucoma in the rare susceptible individual, or aggravate the condition in an otherwise controlled patient. Complications seldom arise, though, if the medication is taken in recommended dosages. One should adhere to the basic principle that medications are meant to treat significant disease or to relieve significant symptoms. No medicine should be used casually or in excess. The side effects of over-the-counter drugs are enumerated on the label or container. Instructions should be checked so that symptoms can be recognized. Prescription medications should be taken only according to directions; any unusual effects or symptoms should be reported immediately to the physician who prescribed the drug. The practice of medicine has been revolutionized by the discovery of drugs that cure some diseases, control others, and alleviate painful or annoying symptoms. The price of relief often includes unavoidable and often unpredictable side effects—too high a price to pay if one does not need the medication to begin with.

This chapter does not pretend to list all diseases of consequence to the eye, nor to enumerate all medicines with ocular side effects. The diseases included are those most commonly encountered by ophthalmologists.

12
Refractive Errors

A refractive error in the eye affects its ability to focus clearly the light rays that enter it through the pupil. Except in isolated instances, refractive errors are not considered a true abnormality or disease of the eye. They are part of the normal visual spectrum; in fact, presbyopia, the near-vision difficulty which develops with age, is an entirely normal phenomenon that may indicate a problem or disease when absent. Only extreme refractive errors are occasionally associated with ocular or systemic disease processes or abnormalities. A refractive error is thus only an error in the focusing ability of the eye. Once corrected with a spectacle lens or contact lens, the eye should see normally if no other problems exist. One is not considered blind or legally blind based on uncorrected vision. As long as vision is correctable to normal levels with spectacles, it is considered normal.

It is worth emphasizing that most refractive errors are normal. The so-called simple refractive errors represent about 98 percent of the total and are felt to arise from a chance arrangement of the components of the optical system, each of which is within the normal range itself. The components of the eye that determine its final focusing power are the cornea, lens, total axial length of the eyeball, and depth of the anterior chamber.

Refractive errors, for the most part within the normal spectrum, are thus remarkably common. One study of individuals in the twenty to thirty age group found 27 percent nearsighted and 56 percent farsighted, with or without some astigmatism, and 2 percent with primarily astigmatism. Only 14 to 15 percent were normal. The incidence of nearsightedness and farsightedness does vary with race and genetic background. And of course such statistics will vary depending on what range one defines as normal.

The hereditary aspects of refractive errors are complicated by the fact that no single factor determines the refractive state of the eye, as we have seen. For the most part, however, the simple refractive errors seem to follow a basically dominant mode of inheritance. Dominant inheritance means that one affected parent carrying a single dominant gene would pass on the characteristic to half of the offspring; two affected parents will pass it on to three-fourths. If a parent has a double dominant gene all of the offspring will exhibit that characteristic. For refractive errors, these are just estimates. Multiple factors determine the final refractive state.

The severe and pathological refractive states are fortunately more likely to be inherited recessively. Thus one affected parent will pass the gene along to his offspring, who will be carriers but not develop the problem unless the other unaffected parent is a carrier. If this unaffected parent proves to be a carrier of a recessive gene, half the offspring will be affected. And if two affected individuals mate, all offspring will be affected. Again, these rules are for single-gene, simple traits and only approximations for the multifactorial inheritance of refractive errors.

The power of any optical system such as the eye is described in terms of its ability to gather and focus light rays. We see because objects either emit or reflect light rays, which enter the eye, go through the pupil, and strike the retina. Each point of any object in our field of vision acts like a point source of light giving off rays in all directions. For practical purposes, once such a point is eighteen to twenty feet or more away from us, all rays that come from it and hit the eye are essentially parallel (figure 12–1). The eye must take these parallel light rays and bring them to a clear focus again on the retina.

The refraction of the eye for distance determines the spectacle lens necessary to accomplish this clear focusing if the unaided eye

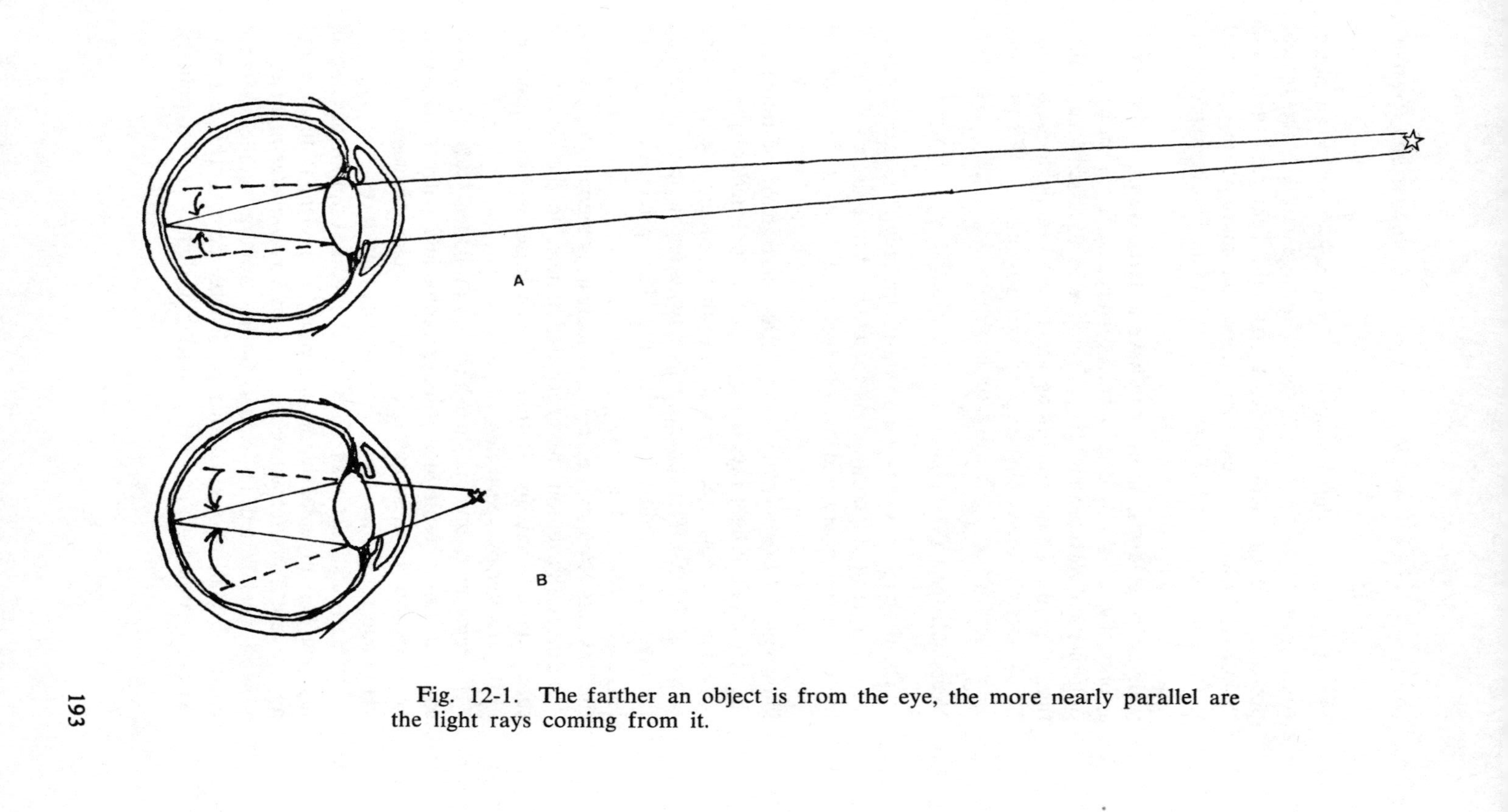

Fig. 12-1. The farther an object is from the eye, the more nearly parallel are the light rays coming from it.

cannot do so. As demonstrated in figure 12–1, when an object comes closer, the light rays from it diverge more. Since optical power is defined as the ability to take the light rays and bring them back down to a clear focus, more and more power is needed to focus on an object as it comes closer and its light rays spread out more when they enter the eye.

In simplest terms, an eye can have a refractive error because it is optically too strong or too weak, or because it has an irregular, nonspherical surface that brings light rays to a blurred focus rather than a point. These conditions are called nearsightedness (myopia), farsightedness (hyperopia), and astigmatism. We will consider each of these refractive states separately.

Nearsightedness (Myopia)

The nearsighted eye is too powerful optically. The excess power may be a result of a cornea-lens combination that gathers light rays more than necessary and brings them to a focus in front of the retina. Or it may come from an eyeball that is too long, a disorder known as *axial myopia.* The focus is also in front of the retina and the refraction of light rays is too great for the length of the eye.

Myopia is corrected with a lens that reduces the refractive power of the eye. The concave lens spreads out rather than gathers the rays, so it is called a "minus" lens. The profile of this type of lens and its function in a myopic eye are shown in figure 12–2. As the lens gets stronger, i.e., corrects increasing amounts of myopia, it becomes thicker on the edges and thinner in the center. The safety and cosmetic implications of such lenses will be considered in the chapter on spectacles.

Extreme axial myopia, in which the eyeball is truly elongated, can sometimes result in a thinner retina and patches of retinal degeneration. In these extreme cases, macular problems may develop and retinal tears and detachments may form.

An early cataract may have as its initial symptom developing myopia. This myopia, due to hardening of the lens in the eye and an increase in its refractive index, is called *lenticular myopia.*

Most nearsighted individuals suffer from a simple nuisance condition that follows a rather predictable course. This common type of myopia usually develops during or slightly prior to puberty.

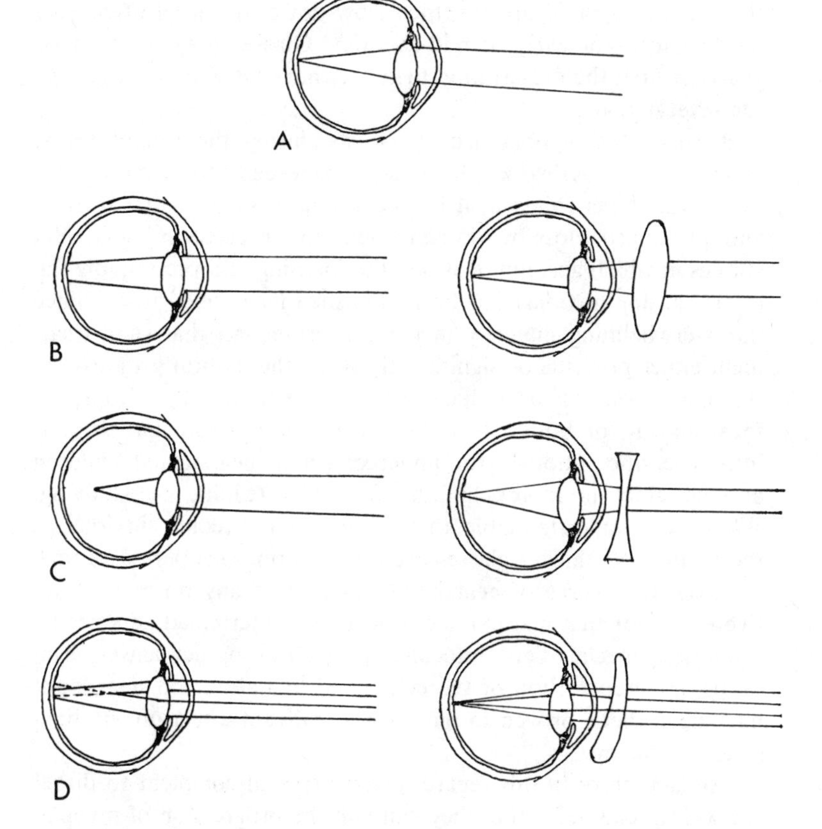

Fig. 12-2. At the top (A) is a diagram of a normal-sighted eye focusing parallel light rays from a distant object on the retina. Since a farsighted eye (B) is not powerful enough to focus these light rays on the retina, the focus would theoretically fall behind the retina. In this case, a convex, condensing "plus" lens is used to focus the light rays on the retina. A nearsighted eye has its focus in front of the retina. A concave, "minus" lens is used to reduce the power of the myopic eye and allow the focus to fall back on the retina (C). An eye with astigmatism has different focusing powers in the different meridians. A cylindrical lens equalizes these powers and the focus is adjusted onto the retina with combined spheres if necessary (D).

It may progress with fairly rapid and significant changes during these years of rapid growth, then slow down and level off during the late teens or early to mid-twenties. Glasses may be changed yearly at first; the prescription then may have no significant change for several years.

Because this myopia usually begins during the school years, when a lot of close work is being done, and seems to be more common in studious children, it has been long suspected that reading and prolonged close work contribute to or cause myopia. Yet studies in which accommodation (the focusing and presumably the effort causing myopia) has been abolished have failed to convince the overwhelming majority in the eye profession that such treatment either prevents or significantly slows the natural progression. Accomodation can be reduced or eliminated with eye drops, bifocal glasses, or both. Most are still of the opinion that there is little scientific rationale for undercorrecting nearsighted children and certainly none for limiting a child's reading or studying. While it is understandable that parents will become anxious as their child gets thicker glasses each year during this period of progression, the degree of nearsightedness (or of any normal refractive error, for that matter) seems to be predetermined genetically. Exercises, special diets, bifocals, spectacles, contact lenses, eyedrops, undercorrection of the error, and limitations on near work have never been proven to have any significant effect on the final state of the eye.

Contact lenses in this regard deserve special comment to dispel the widespread belief that they can halt the progression of myopia. It is simply not true. Contact lenses should float on a layer of tears over the eye. They do not shape the eye or restrict any changes in it. If contacts fit so that they actually alter the correct curvature, they can cause problems, including permanent corneal scarring and irreversible loss of vision.

The belief that contacts slow the progression of myopia comes from three coincidental factors. Most contacts are not prescribed until a child is at a responsible age and able to handle the lenses intelligently. This age, usually the mid- to late teens, often coincides with the time that the eyes are stabilizing anyway. Second, for significant, large refractive errors the change in a corrective lens that compensates for a given change in the eye is actually smaller the closer the lens is to the eye. Thus, the same change in the eye will require a smaller change of correction in a contact lens than in a spectacle lens. Finally, since patients cannot see the thickness of the contact lens, they are less likely to make alarming comparisons to previous lenses, as they would with spectacles. On the whole, contact lenses may be desirable from a cosmetic or optical standpoint, but if fit properly they will do nothing to alter the physical state of the eye itself or its natural tendency to progress or stabilize.

Most eye doctors agree that a child should be given the best correction possible for myopia. No solid proof exists to warrant partial correction or other variably disabling and disrupting regimens in an effort to reduce the final extent of the nearsightedness. Very seldom does nearsightedness constitute more than the nuisance of needing glasses for clear distance vision. Spectacles are not only attractive and an accepted fact of life, but today's hardened and plastic lenses actually protect against flying objects and accidental blows to the eye. A greater array of contact lenses, which can be fit to increasing numbers of individuals, is now available. Those who prefer them to spectacles for cosmetic reasons or because of activities that limit the use of spectacles now stand an excellent chance of being fitted satisfactorily.

Farsightedness (Hyperopia)

Farsightedness is the opposite of nearsightedness. The farsighted eye is too weak optically. It cannot gather the light rays entering it and bring them to a focus on the retina. Because this gathering power is insufficient, the focus falls behind the eyeball (figure 12–2). Either the eyeball is physically too short for the power of the cornea-lens combination, or the cornea-lens combination is simply too weak optically. As with myopia, most farsighted individuals fall within a normal spectrum. Only unusual and extreme instances are cause for concern. As discussed in the chapter on strabismus, extreme farsightedness in children may lead to crossed eyes and then amblyopia. The farsightedness must then be corrected to eliminate the crossing. In other rare instances, the farsightedness may be associated with an abnormally small eye or shallow anterior chamber. As explained in chapter 5, a shallow anterior chamber may indicate a predisposition to acute glaucoma. Eye doctors always consider these associations in evaluating a farsighted patient. As noted, though, hyperopic individuals usually need nothing more than glasses.

The average normal infant is born farsighted. As the child ages, the hyperopia diminishes, most markedly during puberty when those who started life without this hyperopia become nearsighted. Hyperopic individuals left with some farsightedness after the growth period will become aware of it at a variable time later in life depending on the extent of the hyperopia and the visual demands of their occupation and life style.

In chapter 3 we discussed the development of the lens in the eye. With age, this lens gradually loses its ability to contract into a more convex shape and thus provide more optical power. The amount of focusing power present at any given time is a remarkably reliable index of age. Infants and small children have a tremendous capacity; by age eighty most of the elasticity of the lens is gone. The practical reflection of this focusing ability is the range of vision. A small child can focus on an object at extremely close range; the distance needed to focus clearly increases gradually with age.

Recall that objects in the distance send essentially parallel light rays to the eye, which then are focused on the retina. As an object moves closer to the eye, the light rays coming from the object hit the eye with a larger angle of divergence and more focusing power is then necessary to gather the rays to a point on the retina. This focusing power is provided by the lens, which contracts into a more convex shape in response to the contraction of the ciliary muscle in the eye. Contracting this muscle in the eye actually loosens the suspensory fibers of the lens, which then allows the lens to contract down into a more convex shape. As the lens loses its elasticity with age, it responds less to the loosening of these fibers and provides progressively less power. Figure 12–3 depicts the focusing ability of the infant and the regular diminution of the ability with age.

With this background information, one can now understand how hyperopia can be compensated for by focusing the lens within the

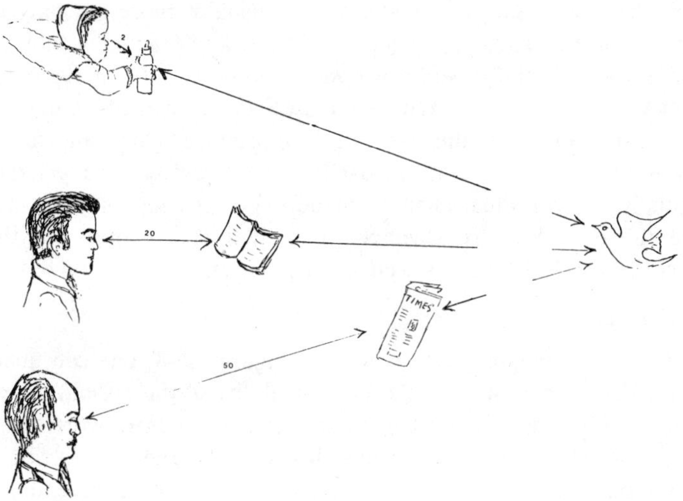

Fig. 12-3. A steady diminution in focusing ability of the normal eye occurs with age as the lens in the eye becomes less flexible. A young child can resolve detail a couple of inches from the eye. By the mid-forties the normal-sighted eye cannot focus to resolve this detail much closer than about 20 inches. By age 80, this near point has receded to about 50 inches. Distant objects will still be seen well, however, as distance acuity reflects the relaxed, least convex, and least powerful conformation of the lens in the normal eye.

eye. Unless it is extreme and gives rise to an esotropia, hyperopia is easily adapted to. Problems arise when visual demands or age make focusing a conscious effort.

Because near work demands focusing, it is during reading and other close visual activities that the hyperopic eye must do most of its focusing. The first symptoms of decompensating hyperopia are eye strain, headaches, blurring, or other nonspecific complaints during or after close work or reading. Glasses can be given to hyperopic individuals when symptoms and circumstances warrant them, but these people will rarely accept a full correction for full-time wear. They will initially use the glasses only to relieve the symptoms that occur during near work. The constant state of focusing that has been maintained for years cannot be released completely and quickly. These glasses usually represent only the correction for distance, or a part of it, and will actually blur distance vision initially until the *tonic accommodation* of the ciliary muscle and lens is relaxed. Such accommodative effort does no physical damage to the eye but can produce symptoms of eye strain. Often, then, only a partial correction can be tolerated initially, with gradual changes made over a period of years until the full correction is accepted or needed. Correction will then provide good distance as well as near vision.

Presbyopia

From the foregoing discussion and figure 12–3, one can understand that sooner or later the focusing ability of the normal eye will reach a point at which it can no longer resolve close work either. The normal-sighted eye reaches this stage in the mid-forties. A common complaint at this age is, "My arms aren't long enough," meaning that reading material can no longer be held far enough away to focus on it. The exact age at which this receding near focal point becomes symptomatic depends on the exact underlying refractive state of the eye and the visual demands placed on it. Avid readers and people engaged in demanding close work will note eye fatigue and symptoms earlier than those who read infrequently and can move the material farther away to resolve it. The limit is the length of one's arm or the size of the material, which may be too small to be seen clearly at any more than a very short distance. Keep in mind that this hardening of the lens and loss of focusing power, which is noted as a difficulty in focusing close up, is a perfectly normal, unavoidable consequence of aging. If it does not develop, there might be a problem.

In a farsighted eye this process is accelerated, so difficulty with near vision will develop before middle age. In a nearsighted eye the difficulties are postponed somewhat, and if the glasses are simply removed the nearsighted eye will still be able to see close up because of its excessive power. The ability to do close work and read easily without glasses will depend on the amount of myopia present and hence the distance at which one can focus.

Glasses are used for correction of presbyopia. Commonly referred to as *reading glasses,* they add power to the eye for near focusing and blur the distance. Because they hamper distance vision, these reading glasses must be removed when looking up and across the room, or even at objects only several feet away. The annoyance of removing the glasses frequently motivates individuals who are constantly looking up and down, such as receptionists and teachers, to get either bifocals or half-glasses. In bifocals the power needed for close work is in the bottom segment, through which one looks automatically when reading and writing. If there is no significant refractive error for distance, half-glasses, the Ben Franklin style, serve the same purpose. The types of bifocals will be discussed in chapter 13. Bifocal contact lenses have not proved acceptable to many people. If conventional contact lenses are used for distance correction, reading glasses are still necessary. If both eyes are corrected for near with contact lenses, one cannot see well in the distance. An exceedingly rare patient has avoided wearing glasses by correcting one eye for near and one eye for distance with contacts. The imbalance is more than most people can tolerate.

Most people need reading glasses by the time they are fifty. Initially, only a relatively weak "plus" lens is needed to add to the focusing power still present. By relaxing the focus of the eyes with these weak near glasses, one can adjust the focus and allow it to recede quite a distance. The focusing capacity of the eye diminishes with age, and stronger reading glasses are necessary to bring the focus still closer. With the stronger glasses on, one can no longer relax the lens of the eye to allow the focus to recede as much as it did before. The symptom noted at this stage is loss of clear vision in the intermediate range, from about 1.5 feet to 10 feet away. The shortening of the range of focus may be quite bothersome to musicians, typists, or others who need to see material several feet away. If one needs clear near vision for fine detail in addition to this intermediate range vision, trifocals can be used; otherwise the power of the bifocal can be compromised. Fine detail will be sacrificed in order to lengthen the range of focus. Separate glasses can be made up to satisfy individual visual needs.

The unfortunate fact of nature is that the normal aging eye gradually loses its focusing power and requires progressively stronger glasses, which in turn draw in the focus and limit the range of clear vision. Figure 12–4 illustrates this normal course of events.

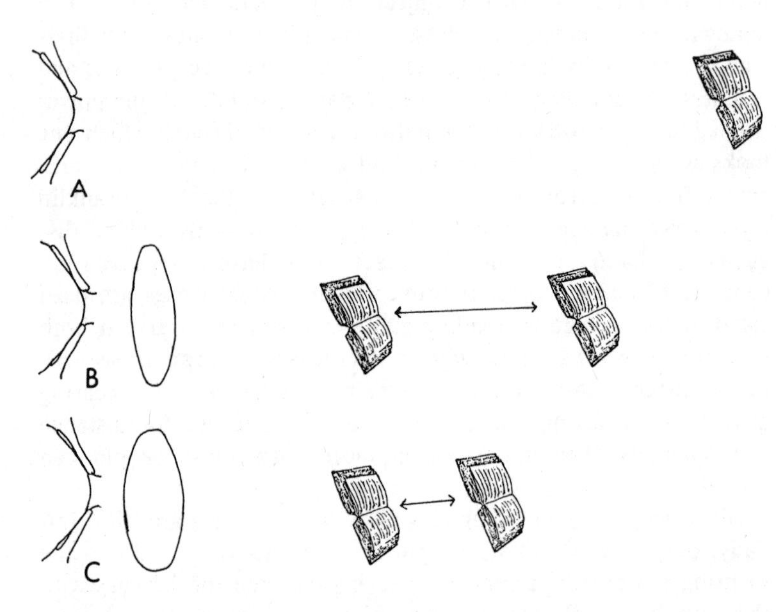

Fig. 12-4. With presbyopia an object can be focused only far from the eye (A). In the early stages a weak convex lens is used to supplement the eye's focusing ability so that the object can be seen at a closer distance. With this lens in front of the eyé and complete relaxation of the eye's focusing, the range of clear near vision can be expanded outward (B). As stronger spectacles are needed to compensate for diminished focusing ability of the eye itself, the range of near vision is necessarily limited. The stronger reading spectacle focuses light from a closer point and no clear vision is possible father out (C). The range of near vision inward from this point represents the residual focusing capacity of the eye.

Astigmatism

The previously described refractive errors represent problems with the overall power of the eye as an optical system. With too little power (hyperopia) or too much (myopia) the focus of light entering the eye falls behind or in front of the retina, respectively. Another type of refractive error affects not the absolute power of focus, but its quality. Only a regular spherical surface (like a portion of a globe or round ball) can bring light rays to a point focus. Recall that light rays emanate from a point in space and then are gathered back down to a similar point when focused by the eye. If the cornea is not spherical, however, the light rays diverging from each point we see in space will not be focused to a similar point on the retina. The best focus that can be obtained is a small blurred circle rather than a point. This type of refractive error, in which the cornea is ovoid rather than spherical, is called *astigmatism*. The shape of the eye will not appear distorted to an observer, but at the very fine microscopic level it can be measured and found to have different curvatures at different points rather than the identical curvatures that characterize a perfect sphere.

Regular astigmatism is the common type and refers to the fact that there is a regular gradation from the point of greatest to least curvature. *Irregular astigmatism* results from disease or trauma and implies that the cornea is quite distorted with no such even gradation from one meridian to the next.

Regular astigmatism can be corrected by a spectacle lens that contains a cylinder. A cylinder lens is ground in the shape of a segment of hollow tube. The smaller the diameter of the tube, the stronger the cylindrical correction. The axis of this cylinder indicates its orientation. It is aligned so as to compensate for an opposite cylindrical formation of the eye. The net effect should be the creation of a neutral sphere that can focus the light rays to a point. If the point so focused does not land exactly on the retina, some hyperopia or myopia coexists, so the appropriate spherical correction is ground into the spectacle lens along with the cylinder (figure 12–2).

In most patients with astigmatism, the axis of the astigmatism is either vertical or horizontal. This orientation is fortunate because it conforms with most contours of the environment and produces the least distortion. The spectacle correction of astigmatism is not perfect, and because of the optical properties of cylinders a certain amount of distortion is unavoidable. Most adults with mild to moderate astigmatism do perfectly well with spectacle correction, as do almost all children even with large cylindrical corrections at an oblique axis. Children adapt readily and gain good acuity with no problem, whereas adults who are given moderate or large astigmatic corrections, especially at oblique axes, will note distortion and often have some problems adapting. It may be necessary to sacrifice some acuity (discrete vision) for a more comfortable partial correction. The patient can gradually build up to a full correction if necessary.

The ideal correction for astigmatism from the optical standpoint is a hard contact lens that will substitute a perfectly spherical front surface for the astigmatic cornea. As pointed out in chapter 14, a soft contact lens will not yield such good results, for it might assume the astigmatic shape of the cornea. If irregular astigmatism exists, as in a scarred or diseased cornea, no spectacle correction will be nearly as good as a contact lens for these same reasons; the irregular corneal surface can be replaced by a perfectly spherical cover lens rather than an approximate cylindrical correction.

The symptoms of astigmatism are poor vision, eye strain, and headache. If the error is large enough, poor vision will be the initial complaint, especially if combined with significant near- or far-sightedness. Small astigmatic errors, however, may still allow good acuity. Eye strain develops with demanding visual work because an accurate focus is not possible, and the eye is trying continuously to find the best focus. Tearing, redness, pain in or about the eyes, headache, or just a tired feeling occur after only a short period of such work. Glasses may be used part-time to relieve or avoid symptoms during visually demanding activities.

In concluding this chapter some general comments about glasses are in order. With the exception of those used to correct a muscle imbalance (see chapter 7), no permanent physical changes occur in the eyes as a result of the glasses. Some amount of focusing or relaxation of focusing the lens in the eye due to the action of the ciliary muscle is natural and has no permanent, harmful effect. Glasses that relax this muscle, or even those that stimulate it, produce no abnormal or permanent change in the eye. Spectacles for correction of a refractive error clarify the focus of the eye, relieve some of the effort of focusing, or both. But this help is optional.

If one does not care to see this well all of the time, no harm is done by not wearing the glasses; nor is there harm in wearing them more often. Patients often report that they wear their glasses only when absolutely necessary because they don't want their eyes to get used to them and become dependent on them. One can get used to seeing well and thus want to wear the glasses more often, but the eyes will not undergo any irreversible physical changes from the spectacles.

Chapter 13 discusses spectacles, types of lenses, types of bifocals and trifocals, and tints. Chapter 14 describes the various types of contact lenses presently available.

13 Spectacles

Glasses are an accepted, routine part of life for many people in our society. Those who wear glasses almost all of their waking hours feel naked or hardly recognize themselves without them. Yet the initial diagnosis of a refractive error is frequently met with dismay and apprehension about having to wear spectacles. There is no need to worry about appearance, though, because a variety of frames and cosmetically appealing designs are now available. In fact, many frames are from collections of famous fashion designers, and they often enhance rather than detract from one's appearance. Functional designs are available as well. Special heavy frames with rubber gaskets to hold the lenses can be worn for contact sports, while wraparound styles and goggle designs protect against work hazards.

Despite caveats about practicality in choosing a frame, appearance almost always dominates the decision. One should be aware, however, that with significant refractive errors the lenses may be thick and will get thicker the larger the eyepiece in the frame. A minimal myopic correction in an average-size frame will have inconspicuous edges usually completely hidden by a plastic frame and still unobtrusive in a metal rim. But in an oversize frame the edge may be considerably thicker and cosmetically unappealing. Some prism effect may be more apparent to someone looking at the wearer from the front. With "minus" concave lenses that correct for nearsightedness, this effect causes the edge of the face to appear displaced inward, indented, when viewed through the spectacle lens. A "plus" convex lens creates the opposite illusion of a bulge in the face when viewed from the front. These effects are minimal with smaller frames and exaggerated with larger.

If the lens size or shape changes in response to a change in frame, the new glasses will usually feel different than the old ones to which one has become accustomed, even if the prescription is exactly the same. One should be prepared to adapt to the new feeling; of course, if true discomfort persists, the prescription must be checked and the fit reevaluated to be certain that the lenses are properly centered and positioned.

Apart from cosmetics, one should consider the anatomy of the face and nose. A very thin, high bridge of the nose, for example, will not comfortably support a frame that rests at this point. Some individuals cannot tolerate the nasal "feet" that provide support for other styles, especially if the spectacles are quite heavy. Whether the earpieces fit around the ear or simply over them should depend on one's need for a stable fit and the sensitivity of the skin around the ears. Opticians can display a variety of suitable frames and discuss advantages and disadvantages of each.

Allergy to the materials in frames is another consideration. Nickel in some metal frames is a common offender; materials or dyes in the plastics sometimes excite skin inflammation and itching. Unless a known history of allergy to such materials exists, only experience will tell. The highly flammable plastic nitrate materials used in some imported frames in the past have been banned in this country now. One need not worry about this problem if spectacles are purchased in the United States, but "bargains" obtained outside this country should be evaluated carefully.

Quality is an overriding factor in choosing a frame. If the spectacles are for occasional close work only, it is obviously less important than if they will be worn almost all of one's waking hours. When one is reliant on glasses, durability is essential. It is usually worth the extra cost and even a compromise on style to avoid the inconvenience of broken earpieces, lenses that fall out, or frames that easily become malaligned. Metal reinforcements in plastic earpieces, well-constructed, reinforced hinges, and eyepieces that can hold the lenses securely are quality features one should look for. The frames should also fit comfortably and remain stable and secure when the head is moving.

Spectacle Lenses

The transparent material from which spectacle lenses are made is important for two reasons. Vision is the most obvious. To provide an undistorted view, the material must be perfectly transparent and free of imperfections. But the material also must be impact resistant. Everyday life, in addition to occupational hazards, presents enough exposure to potential trauma that safety is an important factor.

Even though lenses are ground exactly to specifications, using pure glass or plastic, certain problems are yet to be overcome. The most significant of these is distortion and a prism effect that shifts the apparent position of objects in space if one looks off-center through a high-powered lens. A more detailed discussion of this problem is included in the chapter on cataracts. Cataract glasses are the most common example of very high-powered, thick lenses. To overcome the difficulty that arises from looking through the peripheral areas of the lens, *aspheric* lenses have been devised. They are made by changing the curve of the lens as it moves peripherally. While this design is preferable, it has not eliminated the problem. Another attempt at lessening the problems of peripheral distortion and of lens thickness and weight is the *lenticular* lens. This type of lens simply places a smaller diameter optical lens into a plain glass or plastic carrier that is fit to the frame (figure 13–1). The wearer thus has a smaller lens to look through with a restricted field but gains a lighter-weight, thinner lens that is relatively free from peripheral distortion. With the exception of some specialized lens designs, the larger the diameter of a conventional lens chosen for a given prescription, the thicker and heavier it will be. If the lens is a concave one correcting nearsightedness, the edges become thicker; if it is convex, correcting farsightedness, the center becomes thicker. As noted earlier, this fact should be borne in mind when considering some of the large styles currently in fashion. Unless lenses are very powerful or ex-

Fig. 13-1. High-power, "plus" spectacles used after cataract extraction. Aspheric lenses are in the pair on the left, lenticular lenses on the right-hand pair.

ceptionally large in diameter, however, optical problems are minimal and easily adapted to.

Traditionally, spectacle lenses have been made from *crown glass.* This term implies that the glass has standard optical specifications and transmits light exceptionally well without distortion. But without special treatment it will shatter, like any high-quality glass, on sufficient impact. Because eye injuries are so commonplace and a shattered glass lens such an obvious hazard, recent standards have been adopted in the United States making it mandatory to harden all spectacle lenses sold in this country. By either chemical means or a process of heating and rapidly cooling the lenses, the glass is annealed to make it more impact-resistant. The process also makes the outer layer of glass in the lens more likely to crumble rather than fracture into jagged, sharp pieces on impact. To be sure, these processes are a definite improvement over untreated lenses that are fragile enough to break into sharp fragments with only minimal force. But these lenses are by no means unbreakable. Given sufficient force, and depending on the type and shape of lens, they will still break into sharp pieces of glass that threaten the eye. If any unusually hazardous exposures are anticipated, one should consider safety glasses, plastic lenses, or a shield around the spectacles.

Safety Glasses

Safety glass is made by a laminating process that reduces the tendency to shatter on impact. If hit hard enough, the layers of glass fracture but should not splinter into many loose pieces.

Plastic Lenses

Plastic lenses are actually the most impact-resistant and hence the safest of all lenses presently available. Given a sufficient blow to a plastic lens, and it is a considerably harder impact than that required to break hardened glass, the lens breaks into large, blunt-edged pieces. Contrary to some beliefs, plastic is just as good optically as glass and actually has even greater transmission of visible light. It pits less than glass, does not fog as readily, and is significantly lighter in weight. In fact, the only drawbacks to plastic lenses are the potential for slight warpage if tightened too vigorously in the frame or exposed to very extreme temperatures, and the tendency to scratch more easily than glass. Because of their greater safety, many eye doctors prescribe plastic lenses for children and individuals engaging in sports or potentially hazardous activities. The lighter weight is also an enormous advantage for patients with high-powered, thick, and otherwise heavy lenses. A new coating process for plastic lenses seems to have substantially reduced their tendency to scratch.

Tinted Lenses and Sunglasses

There is great interest in tinted lenses in spectacles, for both cosmetic and practical reasons. Undoubtedly, the real need for such lenses is greatly exaggerated and their usefulness overrated. But many people with no particular eye disease to cause unusual sensitivity to light are nonetheless bothered by even moderate glare or light. Other people have occupational exposures to certain lights that can actually damage the eye if protective lenses are not worn. And many simply like the image created by the tinted lenses. Whatever the needs or motivations, some basic facts about these lenses will dispel the many prevailing misconceptions.

No harm can be done to the normal eye from exposure to ordinary sunlight. Our ancestors suffered no ill effects from being without sunglasses, but no harm can come to the eyes from using them either. Visual acuity will be reduced with such lenses, for they cut down the amount of light entering the eye, but if one can accept this, no other problems ensue. Heavily tinted lenses should not be used at night, when not enough light is available for safe vision, but no permanent physical changes will occur in the eye even if one strains to see through the lenses in dimly lit surroundings.

The usual tints in spectacle lenses cut down on light transmission by about 20 percent. This slight tinting may prove helpful to some individuals working under excessive light, but its value in reducing light intensity is minimal. Dark sunglasses can cut light transmission as much as 80 percent.

Many people feel that color is important in sunglasses or tinted everyday spectacles. Unless exposures such as carbon arc, electrical, or acetylene welding; laser work; or glass blowing and open furnace work dictate specific industrial tints to eliminate harmful excessive wavelengths, the color seems of little consequence in practical experience.

All glass absorptive lenses, even untinted, absorb most ultraviolet rays. Green lenses absorb infrared as well, and approximate the natural sensitivity of the eye to various wavelengths (colors). Only neutral gray tints reduce light intensity without altering one's color perception. All other colors affect the spectrum of light selectively and change the apparent color of some shades. Thus, gray and green are the most common colors for sunglasses. Brown absorbs blue light, so blues will appear muted and neutral. Yellow lenses minimize haze by filtering out blue and ultraviolet light, but they transmit nearly 80 percent of the light intensity and have no effect on glare.

Polarized Lenses. Polarized lenses are most useful for motorists, truck drivers, boating enthusiasts, and others who must cope with the glare of reflected light. Polarized lenses will eliminate glare and cut overall light intensity by about 80 percent. Unless one has prolonged and frequent exposures to reflected light, however, ordinary tinted sunglasses that cut light transmission by an equivalent amount usually will be just as comfortable.

Photochromic Lenses. The latest innovation in tinted lenses is the lens that changes color density spontaneously upon exposure to light. These photochromic lenses are impregnated with a silver halide crystal that undergoes a reversible decomposition upon exposure to ultraviolet light. Presently the lenses are available in gray, brown, and in light and dark tints. The lighter tints are useful indoors, where they maintain minimal color and filter out about 15 percent of the light. Outdoors they darken and filter out about 60 percent of the light. The darker tints available as sunglasses filter out almost 80 percent outdoors and about 30 percent indoors. The practical effect of these changes is that an individual who is very sensitive to light will find that the lighter shades suitable for indoors will not get dark enough outside for comfort. The darker shades do not get light enough indoors to suffice for most inside activities, especially in dimly lit surroundings. But many people are very happy with these glasses and find that the darkening effect of the lighter shades is adequate to eliminate the need for an extra pair of sunglasses under most circumstances.

Photochromic lenses are available in glass only, but bifocals and trifocals can be made from it. Individuals accustomed to lightweight plastic lenses should consider the probable discomfort of switching to heavier glass if they are interested in these lenses.

Bifocal and Trifocal Lenses

For the most part, the type and size of bifocal segments are determined by the doctor's prescription and the patient's needs. Whether one has a flat-top segment or a round-top segment is best left to the optician to decide, depending on the prescription (figures 13–2, 13–3). "Image jump," whereby the relative position of objects seems to shift suddenly as one changes gaze from up or ahead to down and through the segment, will be reduced if the style fits the basic curve and optical dimensions of the overall lens. The bifocal segment is placed in the spectacles to allow one to read at a close range. The focus through this segment is thus

Fig. 13-2. A typical flat-top bifocal.

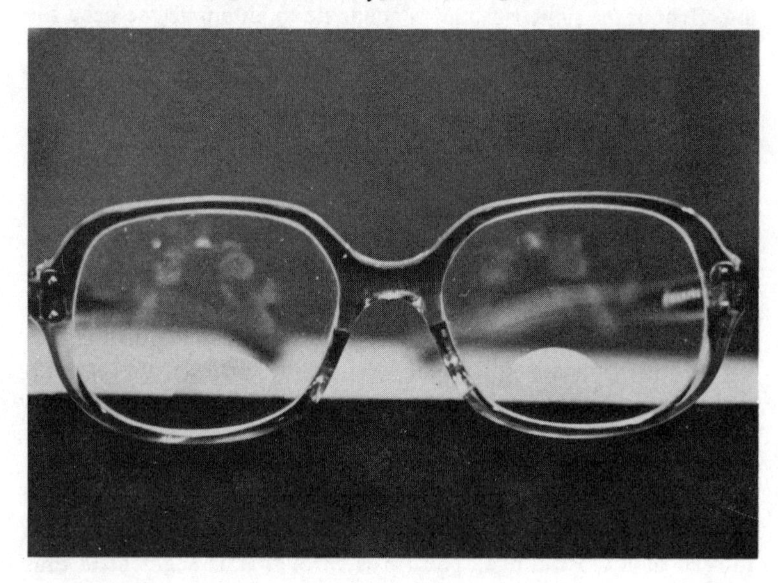

Fig. 13-3. Round-top bifocals.

close up, not beyond arm's length and down on the ground. Stairs, curbs, sidewalks, floors, and all their possible obstacles will be out of focus when one glances down and winds up looking through these segments. As one ages and this segment is necessarily strengthened, the focus comes closer and creates a larger intermediate range of blurred vision. Trifocals are the answer for those who need clear vision at arm's length as well as at close range (see chaper 12). Secretaries, draftsmen, and musicians most often need this range of vision (figure 13–4).

If clear vision down to ground level is necessary, an elevated segment that leaves a band of distance lens below it may be satisfactory. This type of bifocal is called a ribbon segment. Small diameter segments usually allow enough visualization down to constitute little practical annoyance, but if one needs the glasses for reading and close work, especially if one scans a lot of material spread out on a desk, the larger segments or even an executive-style segment may be in order. These cover the entire width of the lens (figure 13–5). Another design adapted for specific endeavors is a lens with the segment located at the top. Electricians, carpenters, and billiard players need such glasses to see at close range while looking up.

Relatively new is the so-called invisible bifocal. Primarily for cosmetic reasons, a bifocal lens was designed which blends the transition zone from upper to lower segment so that it is not apparent to others looking at the lens. The main problem with this lens is that this transition zone is apparent to the wearer as a rather wide blur zone. Another advent is the bifocal lens that attempts to achieve a real visual transition from distance to near vision by grinding a series of intermediate powers between the centers of the distance lens and the near-vision segment. No line or segment is apparent in looking at this lens either. But the optics involved are extremely complicated and not yet mastered. Some distortion is still evident with this lens, and it is unlikely that anyone accustomed to regular bifocals will adapt to these lenses and accept the optical imperfections. New bifocal wearers, however, may find the lenses exciting and acceptable. These graduated bifocals cost more than the conventional bifocals.

Most important in bifocals and trifocals is proper fit and align-

Fig. 13-4. One style of trifocal spectacles.

Fig. 13-5. Executive-style bifocal segments.

ment of the segments. One should not have to contort the head and neck to read or to see over the segments. The segments should be at such a height that a comfortable glance downward brings close material into focus, while a natural head position and gaze straight ahead allow unobstructed distance vision. Obviously, stable, carefully fit frames are as important as the bifocal height, style, and alignment.

Sometimes the doctor misunderstands the patient's real visual needs, and sometimes the patient chooses inappropriate frames or styles. All reputable eye doctors and dispensers of eye glasses verify their prescription and check the fit and alignment of any spectacles they prescribe. Even if an error is not due to doctor or manufacturer, many will make changes at no charge if the problem is brought to their attention in a reasonable period of time. All new spectacles, even if no change in prescription is made, may feel odd at first. If one cannot adapt comfortably to a new pair of glasses in a week, one should return to the dispenser where the spectacles were bought to be sure that the manufacture and measurements are correct.

Subtle changes in manufacture, lens blanks, and alignment, not to mention major changes in style and lens size, may be just as uncomfortable initially as a major change in the basic prescription. One should be reassured by the doctor that the discomfort is a symptom of adaptation rather than the result of a miscalculation.

Most glasses are readily accepted, and the overall quality of spectacles produced in the developed countries is surprisingly good. When one pays high prices for these goods and services, one can expect quality and dedication to patient satisfaction.

14
Contact Lenses

Contact lenses are extremely popular and helpful optical devices for people who need but do not like to wear spectacles. In some instances contact lenses may provide better vision than spectacles; in others the acuity is somewhat reduced. It is far more natural to have the optical correction float on the front surface of the eye and cover the visual axis rather than sit half an inch out in front of it, as a spectacle lens does. A contact lens provides full, undistorted visual field. There are no frames to fall off or slide down the nose, no spectacle lenses to get foggy, and no impediments to physical activities. Apart from the practical and optical advantages of contact lenses, the most common reason for their use is cosmetic. Many people simply do not like their appearance with spectacles. Contact lenses are not for everyone, though. Despite the incredible explosion of new developments and materials in the contact lens field, certain definite precautions and contraindications exist for all lenses. All patients are not suited to all types of lenses, and some are not suited to any. This chapter will discuss the basic considerations involved in deciding to try contact lenses. It will describe the different types of lenses presently available and those likely to be available soon and offer some guidelines for proper use of the lenses.

What Is a Contact Lens?

A contact lens is an optical lens designed to float on a thin layer of tears over the cornea. Only certain therapeutic or investigational lenses are fit tightly in order to protect the eye or alter its natural shape. Correctly fitted lenses float over a certain limited area with blinking and motion of the eyes, thus allowing normal circulation of the tears and gases (oxygen and carbon dioxide) necessary for the nutrition of the cornea. If a lens is too tight and does not move, it deprives the cornea of its nutrients and causes corneal swelling, death of corneal cells, pain, blurred vision, halos around lights, and, in the extreme, permanent corneal scarring and distortion. If a lens is too loose, it drifts out of place and often out of the eye, causing obvious changes in vision. Fitting of contact lenses is crucial and should be done only by a qualified professional who insists upon careful, periodic checkups to assure continued good fit without complications.

Contact lenses first appeared about five hundred years ago, but they did not become truly practical until the 1940s, when lightweight, optically precise plastic was utilized for their manufacture. The glass contacts that had been made in the past by the expert glassblowers and opticians of Europe were too heavy and cumbersome for general use. They served primarily to protect the eyes of patients suffering from severe lid diseases that imperiled the cornea. Polymethylmethacrylate provided material for a lightweight, optically excellent, efficiently manufactured, and easily reproduced contact lens that could be widely tolerated for cosmetic use. Until recently, this material constituted the only available contact lens—the hard lens. Thanks to newer materials, there are now several types of soft, semisoft, and new hard lenses that have wider applicability. Their differences and relative advantages will be reviewed later, after we discuss indications and contraindications for contact lenses in general.

Who Should Wear Contact Lenses?

It is estimated that approximately 90 percent of those who want to wear contact lenses can obtain a satisfactory fit and tolerate the lenses. Without sufficient motivation, though, one will not weather the discomfort of the initial adjustment to the lenses, clean and care for the lenses, and schedule periodic checkups with the eye doctor. It is a mistake to get contact lenses out of curiosity or because your friends are getting them. People who have no complaints about their appearance or vision with spectacles, and who in fact may feel unprotected without them, are not likely to have sufficient motivation to become successful contact lens wearers. And there is no reason they should. Believe it or not, eye doctors get occasional inquiries about contact lenses from patients who do not need any significant visual correction. If one wears spectacles only occasionally and sees well enough the rest of the time, there is really no need for a full-time commitment to contact lenses. A need for greater acuity and definite dissatisfaction with spectacles are the first prerequisites for even considering contact lenses. One then has reason to consult an eye doctor for a medical evaluation of the eyes.

Anyone with certain diseases of the cornea, abnormally dry eyes, chronic infections of the eyelids, or chronic allergies will probably not be able to wear contact lenses. The solf contacts are more readily tolerated and less irritating than the hard lenses, so they are sometimes acceptable to people with mild allergies or extremely sensitive eyes. But any lens that causes significant irritation, redness, and tearing after a period of adjustment is simply not suitable. No lens should ever be used while the eye is inflamed from an infection, allergy, or other ailment. If a lens wearer develops an eye inflammation, the contacts should be removed and an eye doctor consulted. Contact lenses make the cornea more prone to abrasions and infections, and any coexisting infection of the lids may spawn a corneal ulcer. Allergic or irritative inflammations of the eye also render it more susceptible to infection.

One must take into account not only the aforementioned diseases and allergic problems, but occupational and environmental factors as well. If one works outdoors all day in a windy or very dry environment with sand and debris that might easily get blown in the eye, or in front of a furnace, or in an environment with irritating fumes or gases, one is not likely to tolerate the additional ocular irritant of a contact lens. It is especially irritating and dangerous when a foreign body gets under a contact lens. The contact must be removed immediately if one feels a foreign body in the eye. The eye must then be evaluated before reinserting the lens to be sure that the cornea has not been scratched.

If the eyes are medically suited for contact lenses, and if one is motivated enough to care for the lenses properly and removed from occupational or other hazards that make lenses inadvisable, one is ready to join the estimated 7 million people in this country who currently wear contact lenses.

Types of Contact Lenses

The Hard Lens

Polymethylmethacrylate is the lightweight plastic used for hard contact lenses. Optically it is excellent, providing good visual acuity through a stable, perfectly spherical front surface that neutralizes an astigmatism or irregularity in the patient's cornea. The front of the contact lens becomes the new front of the eye's optical system, "replacing" the cornea (figure 14–1). It will provide better vision than spectacles for individuals with significant astigmatism, or corneal scars and irregularities. Its main disadvantage is that it deprives the cornea underneath it of nutrients and oxygen. To be tolerable, the lens must be fit so that it moves enough to allow circulation of tears, oxygen, and carbon dioxide and small enough to allow rapid diffusion into the area under the center of the lens.

Because the lenses constitute hard foreign bodies, an uncomfortable adjustment period should be anticipated, even with the best fit. Most people suffer tearing and irritation for variable lengths of time, but build up to comfortable full-time wear within several weeks.

Because the lens may be quite small to provide the necessary good fit without complications, some visual problems may ensue when the pupil of the eye enlarges. At night or in dark surroundings the diameter of the pupil may approach that of the contact lens, producing glare and blurring from the light rays that enter around the edges of the lens. If a larger lens is not tolerated and the problem is significant, one may have to consider alternative

Fig. 14-1. A hard contact lens fitting properly over the central cornea.

lens materials, fenestrations (holes) in the lens to aid tear and air circulation, or simply discontinue the lens for nighttime driving and other such activities.

Most people who temporarily discontinue hard contact lenses because of an infection or allergy attack find that they need a period of readjustment before they can get back to full-time wear. The longer the lenses have been off, the longer and more cautious the readaptation period should be.

The plastic material of the hard lenses is quite durable, so they may last for many years. If a lens warps or chips, however, it should be replaced. Minor scratches may be buffed and polished out. The lenses can be disinfected with strong antiseptic solutions without harm to the plastic, and they may be stored in cases wet or dry.

The Soft Lens

Several polymers are used in the manufacture of the soft lens. The main difference between the lenses at this time is their water content, which varies in the presently approved lenses from about 30 percent to 60 percent. These lenses are soft and flexible and have as their main advantage increased comfort. Many people who could not tolerate hard lenses readily accept soft lenses, and patients who begin with soft lenses adapt more quickly and easily than they would with the hard variety. Because these lenses are comfortable and have some permeability to oxygen, larger lenses can be fit and overall wearing time is often extended beyond that of a hard lens (figures 14–2, 14–3, 14–4A, 14–4B). The larger diameter eliminates the glare around the edges of a small, hard lens when the pupil is dilated.

But the soft lens has some disadvantages also. Chief among its problems is the quality of vision through it. Often the acuity is not as sharp as it is through spectacles or hard lenses. The vision is especially poor in individuals with significant astigmatism. Recall that astigmatism refers to the fact that the cornea is not perfectly spherical. A hard lens completely neutralizes this error by providing a new, spherical front surface for the eye. A soft lens, on the other hand, often molds to the shape of the cornea and fails to correct the astigmatic error. Some special soft lenses have a correction for astigmatism ground on the surface of the lens. The lens is so designed that it can be kept in the proper orientation to correct the refractive error.

Soft lenses are not as durable as hard ones. They may be torn by fingernails or foreign bodies, and they may discolor from mucous or protein deposits. The latter problem can be treated with a special enzyme cleaner.

Soft lenses must be stored in special solutions, and they should be disinfected each night either in a special heating unit usually provided with the lenses or in a special chemical disinfectant solution. Although many consider this cleaning procedure to be a great disadvantage of soft lenses, it is an excellent hygienic routine. The lenses, whether hard or soft, are routinely removed and stored somewhere at night anyway, so it is hardly bothersome to drop the

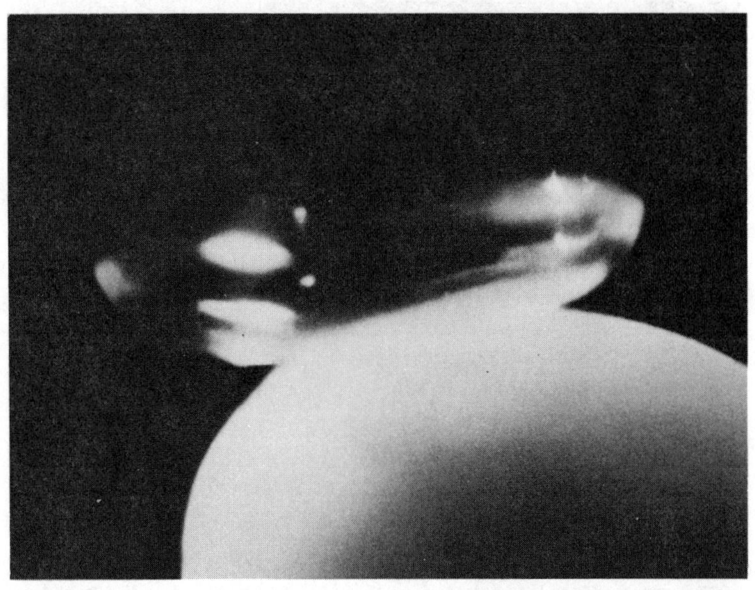

Fig. 14-2. A soft contact lens poised on the fingertip prior to insertion. Its diameter is larger than that of the typical hard lens.

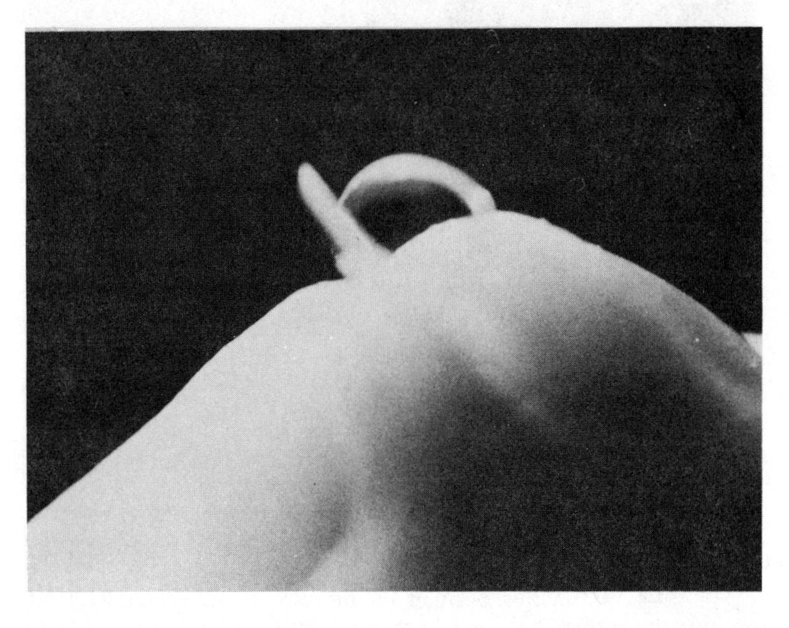

Fig. 14-3. A demonstration of the flexibility of the soft lens.

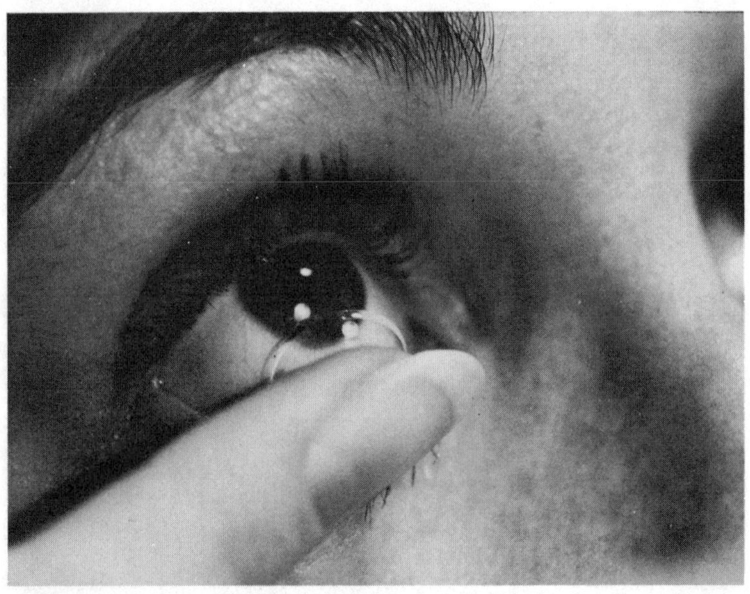

Fig. 14-4. The soft lens is usually placed with the forefinger below the cornea.

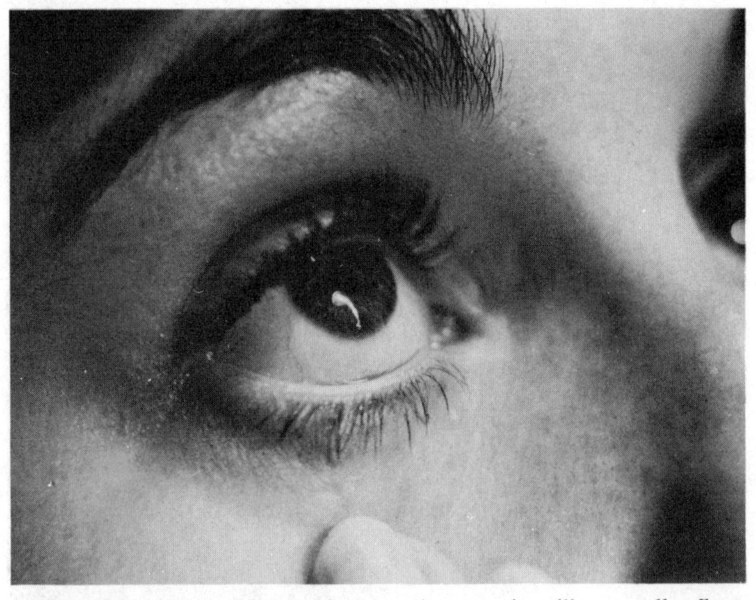

Fig. 14-4. From this position on the eye, it will naturally float up over the cornea. It is usually removed by sliding it down, grasping it with two fingers from the position below the cornea, and folding it out.

storage case into a heater and press a button. The unit goes off automatically and in the morning the lenses are ready for wear. Ready-made solutions are now available for soft lenses so that home preparation of special boiled saltwater from salt tablets is no longer necessary.

More significant is the objection that the lenses are friable and cost more than hard lenses. They need replacement more often, rarely lasting more than several years. The increased comfort of soft lenses, however, has made contact lens wearers out of millions of people who did not want to undertake the more difficult adaptation to hard lenses or simply could not tolerate them.

New Lens Materials

An exciting and wide variety of new plastics is presently under investigation for contact lens use. Some of these materials are not yet approved by the FDA, and are now being used on an investigational basis only until they are proven safe after sufficient long-term use in a large enough sample of people.

Cellulose acetate butyrate is the material used in semisoft lenses. It is actually a firm plastic but, unlike polymethylmethacrylate, it has some permeability to oxygen and carbon dioxide. Hence, these lenses can be worn in a larger diameter and can be tolerated for longer periods of time than the traditional hard lenses. Semisoft lenses are easier to adapt to, and like hard lenses, they will correct astigmatism and provide good vision. They have recently been approved by the F.D.A. for general use.

Silicone is material currently available on an investigational basis. The silicone lens is far more permeable to oxygen and carbon dioxide than any of the other materials presently in use. It holds great promise as a continuous-wear lens. A silicone lens is comfortable, pliable, easy to clean, and affords good visual acuity.

Other variations on the materials already mentioned, such as ultrathin soft lenses, are being developed and evaluated. The ideal contact lens, which still eludes us, is a comfortable, durable lens that provides excellent visual acuity by neutralizing corneal astigmatism and irregularities as well as the basic refractive power defect. It easily covers the visual axis and corneal area over a dilated pupil, is large enough to center well and give a stable fit, and has enough oxygen permeability to disturb corneal metabolism only minimally. Such a lens has not yet been discovered, but the innovations of the past several years represent exciting, giant strides in the right direction.

Full-time Wear Contact Lenses

The implications of full-time wear contact lenses are probably already apparent from the foregoing discussion. No lens presently available can be used without caution on a full-time basis because all contact lenses presently available impede to a greater or lesser extent the normal functioning and metabolism of the cornea. A very small hard lens or a very permeable soft or semisoft lens may well be tolerated full-time for months in some individuals. But anyone endeavoring to wear contacts full-time should be under the care of a responsible eye doctor who will be certain no ill effects develop. Certain soft lenses have recently been approved for extended wear by the FDA.

Bifocal and Multifocal Contact Lenses

Middle-aged and older individuals who need reading glasses or bifocals find that contact lenses fitted for keen distance vision will not allow good near vision. The usual solution is a separate pair of reading glasses worn over the contact lenses for close work. These half-glasses or bifocals with powerless upper segments make no sense for people who need keen near vision most of the time. They might just as well wear glasses alone, unless the contacts provide much better acuity by correcting for extreme refractive errors or irregular astigmatism. People who still resist glasses can try as an alternative a contact lens in one eye to correct for distance vision and a lens in the other to correct for near. Unless one has been adapted by a lifetime of such imbalance, however, this compromise is likely to prove intolerable.

Bifocal and multifocal contact lenses exist in a variety of designs. While in theory they seem as if they should work satisfactorily, in practice none seems to have generated much enthusiasm. Near vision usually is not as clear as one would like, often due to problems with the exacting fit required for these lenses, interference from the tear film, and, in some cases, the insufficient motivation of most older individuals to cope with the inadequacies and annoyances of contacts when spectacles are so much easier.

Warning Signs of Impending Problems or Need for Adjustment of Lenses

Ideally, contact lenses do not disturb the natural functions of the cornea or eye. They simply alter the focus of light rays entering the eye. Of course, the ideal does not take into account a foreign body floating over the major portion of the cornea and interfering with the normal exchange of tears, nutrients, wastes, oxygen, and carbon dioxide. The eye has a great ability to adapt to such alterations, but not recognizing its limitations may be dangerous. An incorrect fit, overwearing, irritation, infection, allergy, or disease may elicit meaningful warning signs. One should remove the lenses at once and consult the eye doctor who fitted them.

If *spectacle blur* appears after one has worn the lenses for a period of time, it may indicate either that the wearer has exceeded his time tolerance for the lenses or that the lenses do not fit as well as they might. The term refers to the fact that one's vision with spectacles after removing the contact lenses is poorer than prior to use of the lenses. The same implication holds true for those whose vision is strikingly better after removing the contacts than it was before use of the lenses. Vision usually reverts slowly to its previous state over a period of minutes or hours. The explanation is that the contact lenses have induced swelling and changes in the corneal shape by altering the normal function of the cornea. People who have worn large contact lenses for many years and neglected such subtle signs may suffer long periods of discomfort and inconvenience. If an infection, injury, or loss of a lens necessitates refitting for spectacles or for new lenses, it may be weeks or months before the corneas actually stabilize and a good, repeatable prescription can be obtained.

If pain, redness, excessive tearing, blurred vision, "halos" around lights, extreme sensitivity to light, or significant discomfort occur, the lenses should be quickly removed. The eye doctor can determine whether the problem is being caused by poor fit, overwearing, an early eye inflammation or disease, a corneal scratch, or a foreign body.

The overwearing syndrome is a common contact lens-induced disorder known to all eye doctors. After wearing the lenses longer than usual, one experiences extreme pain, tearing, poor vision, and exquisite sensitivity to light several hours after removal. This happens because the cornea did not adapt to prolonged oxygen deprivation and some surface epithelial cells broke down. They are sloughed off in the ensuing hours and provoke these dramatic symptoms, which are like those of a welding flash burn or a scratch of the cornea.

In the face of ocular discharge, redness, pain, extreme sensitivity to light, or diminishing and fluctuating vision the lenses should not be used until an eye doctor is consulted to rule out other disease processes and to prescribe necessary medications. Early infections or other eye or lid diseases should not be complicated by contact lenses, which may spread an infection, create a corneal ulcer, or otherwise endanger the eye.

Orthokeratology

This questionable procedure practiced by some optometrists tries to take advantage of the fact that if contact lenses are fitted in such a way that the cornea alters its shape, vision might be improved without the lenses. After wearing a series of lenses that successively alter the cornea, one hopes to achieve clear vision without any lenses or spectacles. Recall, however, that the ideal contact lens does not alter corneal shape or function. Any lens that does is interfering with its normal metabolism. At best these effects on vision are transient; at worst they can result in permanent warping or scarring and loss of vision. While definitive evaluations are not available now, most ophthalmologists are skeptical about orthokeratology because of its potential ill effects.

In summary, contact lenses are an excellent optical alternative for those who are medically suited to them and responsible enough to care for them properly. Never before have so many people been successfully fitted with contacts. The immediate future holds promise for even better lenses with wider application and acceptance.

15
Dyslexia and Reading Problems

Reading is a critical skill in today's society. In fact, many parents look upon it as a key determinant of their child's future achievements in life. If this judgment seems exaggerated and overly simplistic, deficient language and reading abilities constitute at least a significant handicap that cannot be minimized. They deserve to be dealt with rationally and methodically, yet our ignorance of the forms of retarded reading ability has left the door open for emotionalism, well-intentioned though misguided therapy, and frank charlatanism. Children and their parents are sometimes worse off for admirable efforts and justified concern.

Because people often consider reading a function of the eyes and assume that any difficulty must be related to an eye problem, the eye doctor is frequently consulted first. Although this is a reasonable course of action, a significant visual or ocular problem is seldom involved.

Incidence and Perspective

It has been estimated that approximately 10 percent of all school children below the seventh grade in the United States suffer from a significant reading disability. But exactly what level of reading constitutes a disability is defined arbitrarily depending on one's goals and needs. Statistics are gathered on the basis of test results, which means that the tests define the ability or disability. Thus, reading disability is much less common in the higher socioeconomic groups than in the lower socioeconomic areas. Environmental and sociological factors may be part cause and part definition of the problem. A laborer probably will be unconcerned about not having the verbal skills necessary for advanced education. Of course, the overall problem is not as simple as this analysis might imply. We are really discussing an inability to master written language at a functional level.

Dyslexia is an isolated defect in reading and verbal ability, inconsistent with the general intelligence of the individual, who exhibits no other pathology that might cause or contribute to such a problem. Although it is commonly offered as a diagnosis of reading disability, it is only one of many causes of reading disorders. It is the least understood and should be the last diagnosis to be made, the diagnosis of exclusion. Only when all remediable and identifiable factors contributing to impaired reading have been excluded can one conclude that a child indeed suffers from dyslexia.

What, then, are the causes of reading disability? What can be done to correct them? And what can be done if dyslexia is diagnosed? What is the nature of this and other reading disorders? This chapter is an attempt to separate sense from nonsense in answering these questions. It will be a disappointment to parents who are eager to find pat answers or promises of quick and spectacular cures, but better to be disappointed now than after time, effort, and money have been expended in false expectations. The prognosis for children with reading problems is in fact very good—if they are diagnosed early and if they follow through with a rational program of remedial training.

Factors Affecting Reading Ability

Intelligence is the key to verbal ability. It is often difficult for parents to acknowledge that their offspring has no isolated reading disorder but is performing consistent with his intellectual abilities. On the other hand, it is futile, frustrating, and psychologically damaging to force a child to attempt to master a level of ability he simply cannot achieve.

Simple verbal tests of intelligence, however, are not an accurate indicator of the true intellectual abilities of a child with a potential reading problem. The child must also be given a performance I.Q. test. A great discrepancy between the verbal and performance intelligence tests suggests a specific reading disorder.

The role of environment and early intellectual stimulation are critical. It stands to reason that a child in a family with frequently absent or otherwise engaged parents and older siblings will have little or no verbal interchange during his early years. If the parents suffer from a language deficiency, the child will not enter school or a testing situation at the same level as his peers.

Obvious neurological disorders may indicate brain damage which also impairs verbal skills. Seizures; cerebral palsy, paralysis, spasticity, or weakness; involuntary movements; and tremors are examples of such disorders. But such symptoms may well be coincidental to a remediable reading disorder. A neurologist is best suited to make this judgment and advise about reasonable expectations of intellectual ability.

Less definite indicators are the so-called soft neurological signs. These are subtle indications of minimal brain damage, a diagnosis that is often made with reservation. A child who is *hyperkinetic,* i.e., unable to sit still and work or play attentively, who always seems to be in motion with poor concentration, is suspect. He may be emotionally labile as well, changing moods frequently for little or no reason. Given a history of prenatal problems or birth difficulties, the diagnosis becomes more plausible. The significance of such a diagnosis lies in its predictive value for the future intellectual functioning of the child.

Hearing is very important to a child's developing verbal abilities. It is the spoken sound and word that the child associates with what he sees as he learns to read. Thus, a hearing deficiency will create problems with speech and reading. A hearing test should be done in any child who is suspect, as the problem is often correctable.

Emotional disorders that retard reading progress are often difficult to pinpoint. To be sure, a vicious cycle can occur once poor performance has been exhibited in any endeavor. The initial failure or difficulty engenders so much anxiety that repeated attempts are destined to failure. The anxiety may be nurtured by parents who are overly concerned about a child's performance. In these instances the emotional problems are secondary to the initial reading problem, whatever its origin. Primary neuroses or psychiatric disorders that manifest themselves as a reading disorder alone are infrequent. Emotionally disturbed children usually exhibit a broader difficulty with all learning and concentration.

Motivation to read and acquire verbal skills arises both internally and externally from a child's role models and authority figures. It can be absent if the child aspires to no goal requiring verbal ability and is not encouraged to do so by peers and parents.

Visual problems can certainly cause reading difficulties, but they are surprisingly infrequent in patients with a primary complaint of isolated reading disorder. Nonetheless, an eye exam is a good place to start the evaluation. Refractive errors, particularly astigmatism and extreme farsightedness, can render reading and close work difficult. So can intermittent deviations of the eyes and large phorias (latent deviations due to eye muscle imbalance). Difficulty in maintaining the eyes together for normal fusion, to avoid double vision, may produce severe symptoms and reading problems. But a manifest constant deviation of the eyes, a *tropia,* in which one eye is always out of line with the other, with no double vision, is a stable condition that presents no difficulty or strain (see chap. 7). Different refractive errors in the two eyes, especially when great enough to produce a significant disparity in image size produced in the two eyes, are another visual cause of reading difficulties. In fact, any number of eye diseases causing reduced visual acuity can impair reading.

The problems and symptoms produced by these eye disorders are primarily visual and quite different from true dyslexia. Eye problems will cause eye strain, a complaint characterized by headache, pain or tightness in and around the eyes, redness, tearing, fatigue, intolerance to critical visual work, and poor vision. While reading may be the most critical use to which the eyes are put, these symptoms are usually referable to any demanding visual task—night driving, sewing, crafts requiring fine hand-to-eye coordination, bookkeeping, and even television viewing. The problem is not specifically a reading disorder, although reading is hampered. In fact, it has been estimated that even a 50-percent loss of vision will slow reading speed but will not disable reading comprehension. Such is the case with all the preceding visual causes of reading disability. Even severe eye strain from decompensating ocular muscle imbalance will cause only retardation in reading ability, not the more specific reading problem we will now define as dyslexia.

Dyslexia

From the foregoing discussion, one can conclude that dyslexia is not a visual problem of impaired discrimination or acuity. Dyslexic children can see well enough. The problem is an inability to understand and use symbols. They can see and differentiate them visually, but they do not comprehend them as written representations of sounds, of things. Perception, i.e., seeing and discriminating the letters and words, is normal. Conception, the understanding and interpretation of the letters and words, is impaired.

A certain area of the brain is known to govern the interpretation and understanding of visual symbols. This *language center* is usually in the left half of the brain, even in left-handed people. Because motor function (voluntary muscle control) is governed by the opposite side of the brain, much has been made of the fact that many dyslexics exhibit *mixed dominance,* meaning that if one is not right handed, right footed, right eyed, and right whatever, the left hemisphere of the brain is not the totally dominant side. The right side dominates whatever function one prefers to perform with a left-sided appendage or organ. We will examine this issue and its therapeutic implications later. Suffice it to say now that once we have ascended to this cortical level of the brain we are in an area of such complexity of interactions, interconnections, and advanced functioning that our present scientific understanding is severely lacking. We know that strokes, tumors, and other diseases affecting the language center in the angular gyrus of the dominant (almost always left) hemisphere will produce aphasia, a difficulty or inability to handle symbols, which is comparable to dyslexia. We do not know that there is a localized defect in that area in dyslexic children.

Dyslexic children, then, have an isolated symbolization problem. They have normal intelligence by performance testing and show no evidence of neurological disease or brain damage that could give rise to dyslexic symptoms secondarily. Interestingly enough, true dyslexics are predominantly male and more often left-handed than the general population. They also exhibit less well defined overall laterality, a preference for one side over the other when using one hand or one eye. These facts and reports of families that include many dyslexics have fostered the belief that heredity is a possible factor.

The symptoms of dyslexia are directly referable to learning and reading. A child of normal or superior intelligence will do poorly on a verbal I.Q. test. Reading and verbal ability will be inconsistent with other abilities. Reversals of letters and words are common to all young students, but in the dyslexic child they will persist. BAD will be read or written as DAB, WAS as SAW, and so on. Spelling is generally worse than reading. Arithmetic is usually affected as well, though some dyslexics will compensate and excel in math.

So we do not know the exact cause of dyslexia or the exact location of the problem (though a very likely site exists). Many theories and ideas have been put forth to fill the void in our understanding. Some, unfortunately, have been translated into costly but ineffective treatment regimens.

As mentioned earlier, some investigators feel that eye movements and coordination are at the heart of dyslexia. They prescribe complicated regimens of eye exercises to improve tracking and coordination, and they use impressive devices to diagnose and treat the disorders that are found. To be sure, a decompensating eye muscle imbalance will cause visual symptoms and reading difficulties, but these will appear in the form of retarded reading achievement, not true dyslexia. If an ocular muscle problem is the cause of impaired reading ability and delayed progress in school, achievement will improve dramatically once the problem is resolved (see chap. 7). Only rarely are prolonged exercise regimens necessary or advisable for ocular muscle problems.

Convergence insufficiency is a disorder of eye coordination in which the eyes fail to move inward together when focusing on a near object. Eye strain or frank double vision may develop when reading or doing near work. In the vast majority of cases simple exercises help the condition within a few months or less. No complicated apparatus is really necessary, and the exercises may be done at home. If this regimen fails, surgery or glasses with prisms may be considered as a last resort. One must be quite certain of the diagnosis because many factors such as the patient's general health, cooperativeness, and susceptibility to fatigue can affect diagnostic measurements. The eye doctor must evaluate the individual fully before prescribing exercise or other therapy.

Most ophthalmologists will not prescribe glasses or bifocals for children with no significant refractive error or ocular muscle problem simply because the child has reading problems or dyslexia. The psychological advantages of giving a child glasses to "help him to read," when in fact they do nothing physically helpful, are negligible. Psychological problems associated with reading problems are better dealt with directly by a psychiatrist or psychologist. And it is more important to develop a good appraisal of the real nature of the child's problem than to delay with placebos.

In practice, of course, decisions about what constitutes a significant refractive error or muscle imbalance are not always easy to arrive at. A definite gray zone exists because of great individual variability. Some people will find a certain spectacle correction helpful, while others may find it useless. Trial and error is sometimes the only logical course of action, but it can be costly. Parents on a tight budget do not appreciate the well-meaning attempt of an eye doctor to help their child when expensive spectacles remain in the dresser drawer because Johnny feels they make no difference, and his performance in school indicates he is right. The same may be true in the treatment of muscle imbalance.

A study of eye movements in poor readers, both dyslexic and normal, did not reveal enough notable differences to indicate that faulty eye movements cause a reading problem. In fact, the reverse seems to be true. Children with normal reading ability develop the same tracking problems, repetitive motions, and irregular flow characteristic of poor readers when presented with material above their level of comprehension. These abnormal eye movements are probably the effect rather than the cause of poor comprehension.

A similar problem of defining true cause and effect exists

with the theory that overall gross motor coordination is involved in reading ability and symbol interpretation. Again, because poor reading and poor coordination coexist in some children, elaborate programs to develop their overall muscle ability have been devised. Proponents suggest that with improvement of gross motor coordination comes a more definitive concept of self, which may then improve the symbol recognition and fine muscle coordination required in reading. From the knowledge we have of the nervous system, it is difficult for most physicians to accept any causal relationship between motor ability and the intellectual capacity to comprehend symbols. Awkwardness and impaired reading are probably evidence of delayed overall maturation of the nervous system. Any improvement in one probably will be accompanied by an improvement in the other as a coincidence and as a function of the child's pattern of growth. It is unlikely that training to produce premature ability in one area of function of the nervous system will hasten the development of an unrelated and untrained realm of higher functioning.

Laterality is another hotly debated issue. It refers to the definite preference for one hand over the other for manual tasks, one eye over the other for sighting, and one ear over the other for listening on the telephone. The nerve tracks governing these functions cross over to the opposite side of the brain. We see on the left side (each eye's left visual field) with an area in the right side of the brain; we control the movements of the right hand with an area in the left side of the brain. Intellectual functioning, though, is localized differently in the brain. In most people, the area of the brain responsible for symbol recognition and hence reading ability is located in a specific part of the left brain. Approximately 8 percent of the overall population is left-handed. Even in these people, whose right brain is dominant for manual tasks, the left brain controls reading ability. This fact, coupled with the observation that there are proportionately more left-handed individuals in the population of poor readers, has led some investigators to conclude that so-called *mixed laterality* of control by the brain may provoke reading problems. Yet when combined with mixed lateralities of eye, ear, and foot, left-handedness has not proven to be predictive of reading ability in the general population.

There is probably some validity to the theory that left-handers begin their reading and writing training with a handicap. The vectors, or slanting flow of motion, necessary to write all languages, whether they read from left to right, right to left, or vertically, are all designed for the right-handed majority. Reading, spelling, and writing skills are intimately related, so children who can easily follow penmanship instructions will probably pass this hurdle more readily and integrate the feel and motion of the letters and words into their final visual perception of them. It is similar to the auditory process of associating the sounds of letters and words with their visual appearance. Nonetheless, left-handers are certainly adequate, and children should not be forced to change once a preference is established. Only if they are truly ambidexterous or fail to exhibit a definite preference by school age should children be encouraged to use the right hand for writing. The more relevant corollary to reading ability seems to be ill-defined laterality rather than mixed laterality.

A definite preference for one hand and one eye for sighting, etc. emerges normally as the nervous system matures. There is a correlation between no preference and poor reading ability. But again, most neurologists view poor laterality and reading disability as a reflection of the general state of maturity of the nervous system. There is little or no reason to suspect that ill-defined laterality causes reading disability; the level of development of the brain and nervous system is probably responsible for both. If definitive laterality has not developed for specific tasks by the later elementary school years, some generalized neurological problem may exist.

An Approach to the Evaluation of Reading Problems

What, then, should parents do when confronted with the possibility that their child may have a true reading problem? First, they should try to ascertain the scope and nature of the problem both from the school and from home observation of the child. Does the child actually have an isolated reading problem, or is all of his school work faulty? Does he have problems comprehending only written material, or does he do poorly with verbal instruction and performance also? Is his intelligence, as measured by performance testing and estimated by general observation, clearly inconsistent with his poor reading and spelling abilities? Is he tiring from visual work only, from all endeavors, or from none? Are his concentration and attention span normal for his age? All of these questions bear directly on the first issues to be resolved which will define the problem: whether he is in adequate general health, has normal general intelligence, and has an isolated reading problem rather than an overall learning disability. If the parents suspect that the child is suffering from reading or visual difficulties, the eye doctor can perform a simple and harmless examination (see chapter 2) that will save a lot of effort and money if a remediable eye disorder is found. The family doctor or pediatrician should be consulted to be certain the child is in good health, as easy fatigability and subtle illness may impair concentration and intellectual functioning. A hearing test should then be performed. If these evaluations are normal, a neurologist should be consulted to rule out neurological impairment or immaturity. Most children suffering from *maturational lag* are simply undergoing their own individual development at a slower than average rate. They will catch up to the normal achievement level in time. It is important to recognize such children, for unnecessary and frustrating training can produce self-perpetuating emotional problems.

A psychological evaluation can be illuminating. Even if the primary cause of the disability is not emotional, there is almost always emotional overlay to the problem.

But the real task begins once the problem has been defined. Whether additional treatment for a primary cause is indicated or not, a rational program of remedial teaching should be initiated. Particularly in true dyslexia, a systematic, step-by-step approach must be taken. The child should use the auditory, tactile, and visual senses to learn the language. He should learn to write, then read. Word recognition and memorization is not an applicable technique. A word should not elicit a simple "I don't know it" or "I do know it." It should represent a problem to be solved by a systematic, phonetic approach. Feeling, having the child run his fingers in the proper direction over large letters, and hearing, sounding the letter, reinforce the visual image and the concept of a symbol. The normal integration of sounds and letters into words is reinforced by writing.

A teacher's personal attention to the pupil builds rapport and helps the teacher to devise techniques that are suited to the child's needs. The reinforcement gained from success, no matter how simple the task that is mastered, maintains the attitude and momentum that a child needs to continue learning. The school, the pediatrician, and the eye doctor can direct the child to special teachers for remedial reading help. It is important to follow up early a suspicion of a reading disability. A diagnosis of dyslexia made before third grade carries an excellent prognosis. Over 80 percent of these children will be capable of normal work for their grade level after a remedial program. The success rate drops sharply when the diagnosis is reached later. Probably fewer than 15 percent of the children diagnosed as late as sixth or seventh grade are able to perform normal grade work after therapy.

What, then, is the long-term prognosis for the dyslexics? The answer is not clear, for there are as yet no long-term studies addressing the problem. While these people undoubtedly suffer a handicap, some have compensated well enough to become successful engineers; others use memory and ingenuity to circumvent whatever reading or writing is necessary to their fields of endeavor. Hopefully, the future will bring better definition to the problem and more help for the child diagnosed too late for simple remedial efforts.

16 Eye Injuries

Because the eye is well protected by its lids and bony socket, most injuries to it are minor. Unfortunately, however, many rather severe injuries can masquerade initially as only minimal irritation or redness. It is wise to be aware of potential problems and their symptoms and to consult a physician if in doubt.

Foreign Bodies

Cinders, metallic specks and particles, wood chips, plaster, insects, and the proverbial mud are examples of foreign bodies that can get in the eye. When something foreign gets into the eye, it usually elicits a tearing response to wash away the foreign material. Assuming that the material was not grossly contaminated with bacteria or fungi, the natural cleansing mechanism of the eye will be adequate, the discomfort will disappear within a short time, and no other problems will ensue. Contamination from a foreign body or the irritation associated with rubbing and tearing may sometimes result in a mild infection that will become apparent several days later as redness and discharge. The foreign body might rest on or scratch the cornea. Because the cornea has one of the richest nerve supplies in the body, one experiences persistent irritation, redness, and tearing when it is scratched. Oddly, though, such symptoms may not develop until 24 to 48 hours later. A scratched cornea is more prone to infection, which can then develop into a dangerous corneal ulcer. Abrasions require antibiotic treatment and evaluation by a physician.

The foreign body that does not spontaneously wash out of the eye can be more troublesome. It may rest in the inferior cul-de-sac (figure 16–1) down below the cornea, but most commonly it gets

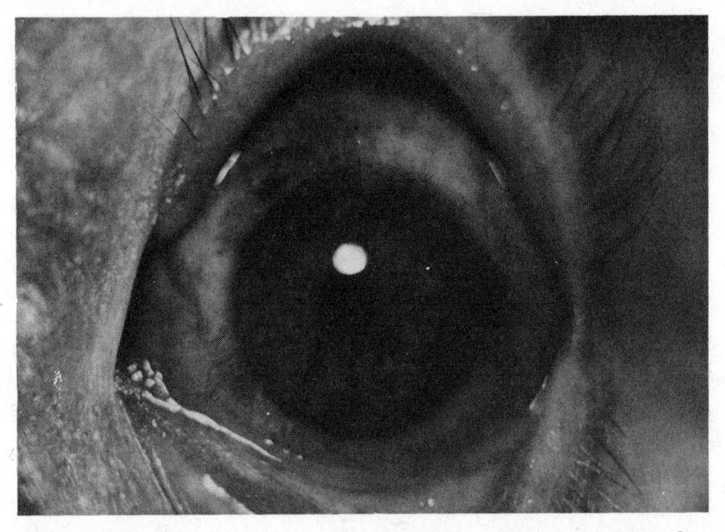

Fig. 16-1. The inferior cul-de-sac is the lower pocket of conjunctiva formed as it folds from the front surface of the eyeball up onto the inside of the lower lid (see fig. 4-4).

under the upper lid. To remove it from this position, it is necessary to *evert the tarsus.* The tarsus is the thick fibrous substance of the lid. It is like a plate that can be flipped outward, exposing its inner surface. The foreign body usually will be seen sitting on the surface and can be wiped away easily. With practice, the maneuver becomes simple and painless. The patient must look down without closing the eyes, while the examiner grasps the upper lashes and pulls out and upward slightly. At the same time a small thin object or finger from the opposite hand pushes *gently* down on a point midway up the lid. If using a pencil or pen, it should always be oriented horizontally so as never to push or poke at the eye itself (figure 16–2). A foreign body found by this technique can be

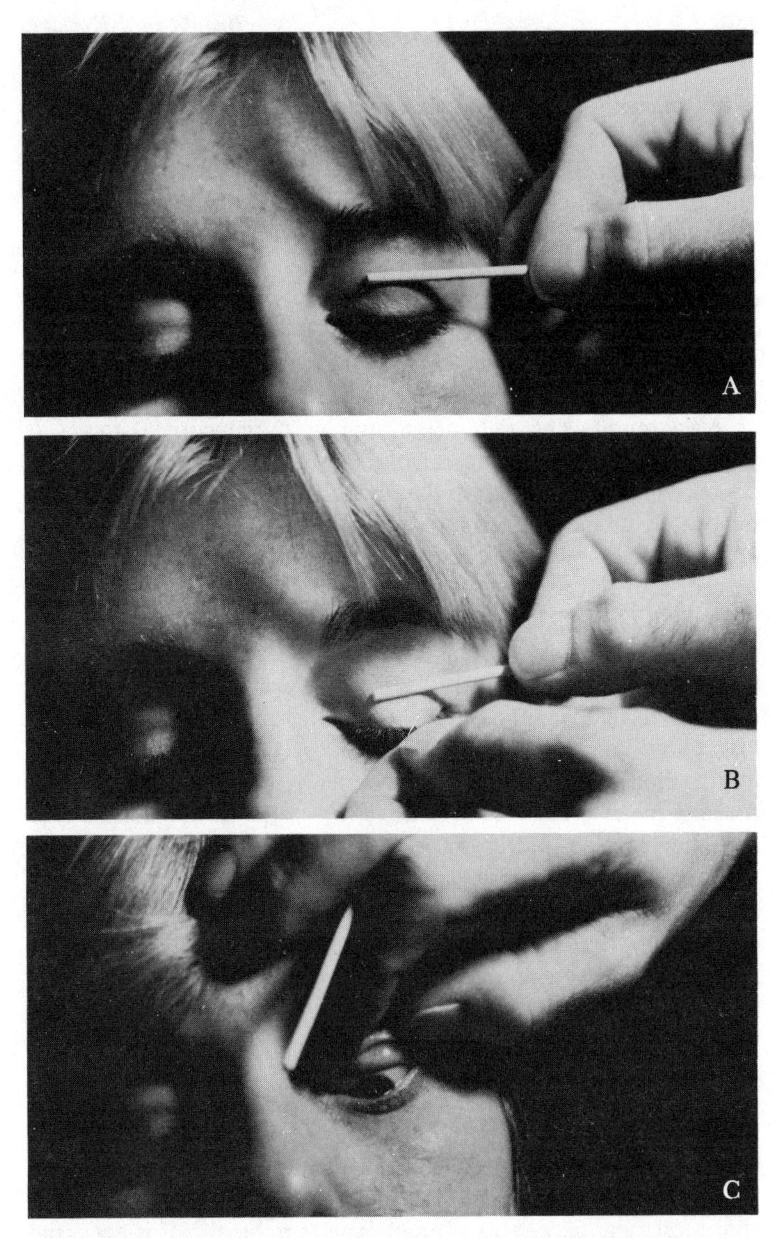

Fig. 16-2. Eversion of the upper tarsus is accomplished by carefully placing a thin object horizontally along the upper lid fold (A), then pulling the lashes outward and upward while moving the object slightly downward (B). The foreign body can be wiped away from everted upper tarsus (C).

wiped off with a clean cotton swab or tissue or flushed off with a stream of water. When the patient looks up, the lid will flip back into normal position. No harm can be done, and the technique may relieve severe discomfort until a physician is reached.

An eye doctor must be consulted if the foreign body works its way higher up into the superior cul-de-sac, for the technique of *double eversion* needed to get at this area is more difficult and requires a special instrument.

A foreign body that remains on the cornea always deserves a physician's attention. It will appear as a spot over the colored iris or black pupil. The rich nerve supply of the cornea makes it painful and futile to try to remove the foreign body oneself. The doctor will instill an eyedrop that numbs the eye and allows careful removal with minimal damage. Metallic foreign bodies are particularly troublesome, as they usually rust rapidly and the rust spreads into the cornea.

Corneal foreign bodies, if superficial, can be removed easily. Healing will progress without scarring or loss of vision if no infection develops. If the foreign body is deeper, however, or if an infection develops into a corneal ulcer, there is danger of scarring, loss of vision, and even loss of the eye. A superficial corneal foreign body may occasionally be removed by irrigating water through the eye, but it *always* leaves a scratch on the cornea that can later develop into an ulcer.

When irrigation is necessary to remove foreign material such as sand particles, it is best carried out with a sterile, balanced salt-water solution or sterile eyewash after the eye has been numbed with a drop of topical anesthetic. Emergency techniques for irrigation of the eyes will be outlined in the section on chemical injuries.

Intraocular Foreign Bodies

A small foreign body does not usually travel with such force and speed that it penetrates the globe. Hammering metal on metal is an act especially likely to produce such an injury, however. A sharp fragment may break off and penetrate the eye. If it does, the victim will experience sudden pain and diminished vision. There may be obvious blood in the anterior chamber of the eye and the pupil may appear distorted. But these injuries can be insidious in that the pain and visual disturbances may be transient. These minimal symptoms are especially likely if the foreign body is small and enters the globe through the sclera, outside the corneal area. It may enter after penetrating through the lid. Sequels to these injuries include infection within the eye, cataract, glaucoma, retinal detachment, and inflammations leading to loss of the eye. Any suspicious injury should be evaluated immediately by an ophthalmologist. Special electromagnets and instruments are available now for removal of such foreign bodies, and prompt surgery has saved many eyes.

Prevention is obviously far better. Protective goggles and safety lenses should always be worn while engaging in activities likely to create foreign bodies or material that can get in the eyes (see chapter 13).

Chemical Injuries

Certain chemicals can cause devastating damage to the eye. Lye and alkaline compounds are the worst offenders. Many industrial cleansers and solvents and most common household drain cleaners, either in liquid or solid form, have a significant enough concentration of alkali to do immediate damage to the eye. These compounds rapidly penetrate the cornea and affect the anterior eye and lens. Corneal opacification, scarring, glaucoma, and cataract may result. Prevention is certainly the best cure, as irreversible harm can be done in seconds. Eye protection should be worn when using these compounds, and even then they should be used with extreme caution. They should never be stored or used within reach of children or placed where they can be tipped over.

Should caustic chemicals or particles get in the eye, it must be irrigated immediately with a copious amount of water. While a sterile, balanced salt solution or eyewash is best, any water in this extreme emergency will do. A hose, faucet, bucket, or anything readily at hand should be utilized. The patient should face the water source but direct his gaze (and thus the cornea) away from the direct stream. Hold the lids open to assure adequate irrigation within the eye. If particles of lye get in the eye, cotton swabs can be used to sweep clean the upper and lower folds of conjunctiva, but care should be taken not to rub the cornea. All of these maneuvers are much easier in the hospital or ophthalmologist's office, where numbing medicine can be instilled first and the proper irrigating materials are available. But one should not hesitate to perform an adequate irrigation with tap water. It may save the eye.

Acid solutions are usually less harmful than alkali because the acid does not penetrate the cornea as readily after it coagulates the protein in the most anterior layers. Of course concentrated acid can still do severe damage, and the treatment is then the same: immediate irrigation, swab away any particles of chemical if it was in solid form, and seek medical attention.

Fumes, sprays, smog, and the like may cause direct eye irritation which is usually minimal and clears within minutes to an hour after exposure. If irritation, redness, or tearing persists, it is wise to have the eyes checked by an ophthalmologist. Irritated eyes are prone to serious infections, but medications are available to relieve the symptoms and prevent infection.

Abrasions

An abrasion is a scraping or rubbing that denudes some of the surface layer. An abrasion of the skin of the lids, for example, might occur after rubbing against a rough surface such as concrete. The skin will appear rough and bleed slightly; if cleaned of all debris, it usually will heal without scarring and without infection. An abrasion of the conjunctiva also will feel rough and probably result in a bright red patch of hemorrhage at the site. While little significant damage is likely, it should be checked by an ophthalmologist.

Corneal abrasions are very common. They can be remarkably painful and often cause tearing and lid swelling. If an object scrapes or rubs across the cornea, the epithelial (outer) cells are damaged and brushed away. If the abrasion is central, vision will be blurred. Even if it is not symptomatic, this type of injury is dangerous because it leaves the cornea prone to infection, ulcer formation, permanent scarring, or even loss of the eye. If no infection or complications develop, healing is remarkably rapid and the epithelium will have regenerated within thirty-six hours with no residual scar or defect. However, constant tearing, pain, irritation, rubbing, and blinking will impede healing. Thus, depending on the severity, the ophthalmologist may place a firm patch on the eye to prevent blinking and promote comfort and healing. Occasionally an abrasion, especially one caused by a fingernail or sharp object, will be deep enough into the outer layers that total, secure healing does not take place. The eye is then prone to recurrent abrasions, or erosions, at the site of the original injury. This problem is discussed in chapter 4.

When a sharp object actually pierces the cornea, the injury is called a corneal laceration. Depending on the depth and extent of the cut into the cornea, topical antibiotics and patching or surgery and suturing of the wound may be necessary. Medical attention should be sought without delay, however, as a partially penetrating or self-sealing penetrating wound might suddenly give way with disastrous consequences. Until a medical evaluation is received, it is wise to avoid squeezing the eye or placing any pressure on it.

Lacerations and Rupture

A sharp object, flying missile, or severe blunt trauma may cut or rupture the cornea or sclera. The injury is immediately painful and causes sudden loss of vision. The laceration or rupture may be obvious or it may occur posteriorly. The pupil may appear distorted and a portion of the colored iris may actually be hanging out of or plugging up an opening in the cornea. The anterior chamber may be so full of blood that details of the iris and pupil are obscured; or the eye may look quite normal from the outside if a rupture occurred posteriorly. If there is any possibility of a ruptured globe, medical attention must be sought immediately. In the meantime, the patient must be advised not to squeeze the lids or move the eyes excessively. Some protection, such as a paper cup, can be placed over the eye with care taken not to place any pressure on the lids or eye itself. Ocular contents can be lost by squeezing, thus making it more difficult to salvage the eye.

Lacerations of the Lids

The first concern of an ophthalmologist in evaluating a cut lid is to ensure that there is no damage to the eyeball. If the cut is at the inner corner of the lids, it may destroy the tear drainage system (figure 3–8). A large deep cut through the upper lid may sever the levator muscle of that lid, causing it to droop. All lacerations of the lids require special care in order to avoid "notching" and deformities afterward. Scarring can lead to contractions and distortions which will require subsequent surgical correction.

When a segment of the lid has actually been torn away, as from a dog bite, the situation is of course more complicated. The eye should be protected and covered if it has been left exposed. Wet clean gauze, cotton, or bandages may be used. If the torn segment can be retrieved, it should be taken to the hospital for possible use in reconstruction.

Blunt Trauma

Severe, blunt blows to the region of the eye can cause many problems. A direct punch or kick, a misdirected tennis ball or elbow, a BB, or even a projectile rubber band are possible causes of trauma. The bony orbit (eye socket) or the eye itself may be affected.

Orbital Fractures

A direct blow on the orbital rims can cause a fracture, as can a blow to the eyeball alone. With a sudden severe blow, the globe expands in height, sometimes enough to break through the very thin bony floor of the orbit. This injury is called an orbital floor, or "blow out," fracture. Its significance lies in the fact that the fracture site may act as a trap door and hold the inferior rectus muscle or its associated tissues. If this muscle becomes so entrapped, the eye will not be able to move correctly and double vision will result. If a huge fracture opens up into the maxillary sinus, which lies beneath the thin orbital floor, so much orbital fat and tissue may fall down that the eye will recede in its socket. For either of these two situations, an operation is necessary, either to free the trapped muscle or to reconstruct the floor of the orbit. If neither complication is present, a simple orbital floor fracture is best left to heal on its own. If such a fracture has other associated facial bone fractures, the need for operation depends on the deformity or functional impairment resulting.

Operations for complicated orbital floor fractures are usually done under general anesthesia and are quite successful as long as the muscle has not been too badly damaged or scarred. In this procedure the inferior eye muscles are freed from any fracture site, bony fragments are removed or realigned, and the orbital floor is reshaped if necessary with a thin inert plastic plate.

The muscles that move the eye may develop traumatic swelling or hemorrhage without a fracture. If scarring does not ensue, these problems usually clear spontaneously and eye movement will be restored.

Traumatic Iritis

Blunt trauma may injure the eye in a number of ways. Besides abrasions and subconjunctival hemorrhages (bright red spots over the white of the eye), it can cause inflammation or hemorrhage within the eye. A *traumatic iritis* is an inflammation within the front of the eye, the anterior chamber. Eyedrops will alleviate symptoms (blurred vision, redness, pain, and sensitivity to light) and prevent scarring of the pupil. If the round muscle of the pupil, the sphincter, is ruptured, the pupil becomes irregular or dilated.

Traumatic Hyphema

Bleeding often occurs in the anterior chamber—a condition called *traumatic hyphema* (figure 16–3). The blood may be diffuse or layered out in front of the colored iris and pupil. Even a tiny amount of blood is significant, as it indicates a ruptured blood vessel within the eye. If the bleeding has stopped, the ruptured blood vessel has been plugged with a clot. But an early clot is not secure. It may open again about three or four days after the injury, when clots normally begin to contract. Rebleeding may then occur, and it is often worse than the original hemorrhage. The incidence of these rebleeds is probably less than 10 percent, but they can be so damaging to the eye that cautious treatment is always warranted. Depending on the circumstances, an ophthalmologist may hospitalize the patient and place patches over both eyes to minimize eye motion; or he may severely restrict the patient's activity at home and check him daily in the office. A large amount of blood in the anterior chamber of the eye can lead to severe

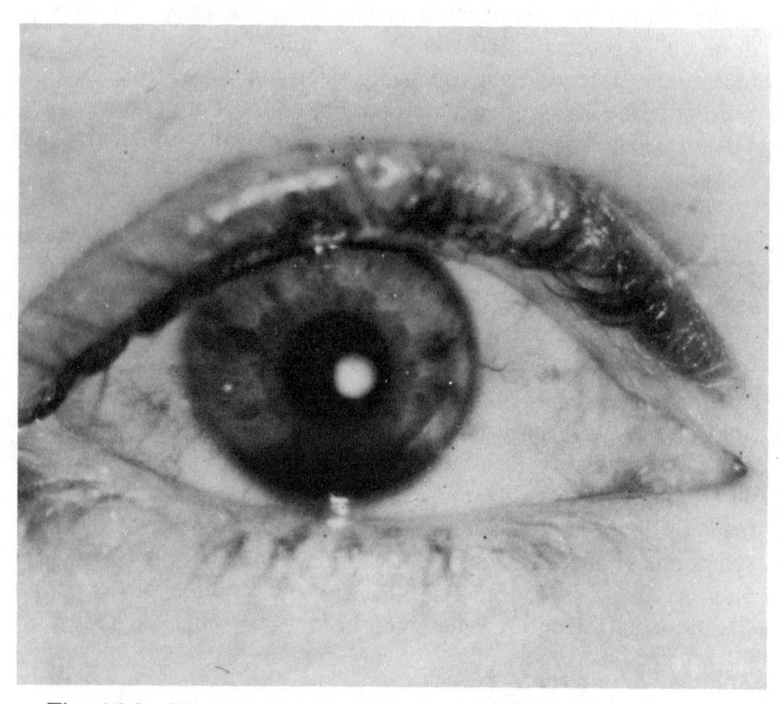

Fig. 16-3. Hyphema is the term for bleeding within the anterior chamber of the eye. Layered blood can be seen below the pupil in this eye previously hit with a ball.

glaucoma and brownish staining of the cornea. Glaucoma may require medicines to reduce the pressure or surgery to remove the clot.

Traumatic Glaucoma

Tearing and scarring of the drainage channels of the eye, the trabecular meshwork, is another result of blunt trauma. Glaucoma can develop from this condition, either immediately or insidiously after several years. It is wise to evaluate this problem initially, even if glaucoma is not present, so that the patient may be forewarned to have periodic checkups lest he develop severe optic nerve damage from glaucoma before realizing it.

Traumatic Cataract, Vitreous and Retinal Hemorrhage, Choroidal and Scleral Rupture.

Cataracts, also sometimes developing much later, can be caused by a nonpenetrating blow to the eye (see chapter 6). And hemorrhage may occur posteriorly in the vitreous cavity. Vitreous hemorrhages must be treated very cautiously, especially since a retinal tear that could lead to retinal detachment is always possible under these circumstances. Minor hemorrhages within the retina, with or without retinal swelling, almost always clear rapidly with little or no subsequent effect on vision. An actual rupture of the choroid, the intermediate layer of the eye, may or may not cause visual problems depending on its location, size, and subsequent scarring. But a rupture of the sclera, the thick white outer coat of the eye, is always a potential disaster. From blunt trauma it may occur just behind the cornea, under the conjunctiva; at the level of the eye's "equator," farther back; or even at the posterior pole. Surgery is necessary to repair these ruptures and any retinal detachment, but the prognosis even with the most successful initial results remains guarded. After such an injury there is a tendency for the vitreous to organize into scar tissue and pull the retina off. Such traction detachments of the retina are quite difficult to repair (see chapter 8).

So the proverbial "black eye" is not necessarily a simple joking matter. Most cases of trauma in or about the eye are easily cared for, but some may be disastrous and require immediate attention. Minor symptoms can mask the true significance of an injury, with grave consequences. When in doubt, it is always wise to consult an ophthalmologist.

17
Resources Available to the Visually Disabled

Visual disability represents a tremendous spectrum ranging from only minimal difficulty in reading and demanding visual tasks to complete blindness with inability to differentiate even light from dark. As we have explained, vision is the total panorama we see—the central keen sight we call visual acuity, plus the expanse of side and peripheral vision known as visual field. A patient with perfect 20/20 visual acuity as tested on the eye chart but severe loss of visual field may be as functionally blind as the individual who can barely see the outlines of the chart. Visual field is necessary for mobility and safety; good visual acuity is necessary for most visual tasks such as driving and reading. Loss of either visual acuity or visual field constitutes a disability.

Arbitrarily, though, blindness has been defined for legal purposes as best corrected vision in the better eye of 20/200 or less, or visual field constriction to less than 20 degrees. An individual just qualifying for this definition will be able to live an independent life, aided by the devices discussed in chapter 9, but will nonetheless be sufficiently disabled to qualify for an income tax deduction and the free services of many government and private agencies and institutions. Lay people often think of blindness as a total inability to see. To be sure, conditions still exist that cause total blindness, and individuals so afflicted are far more disabled than others.

The extent to which an individual copes with the disability, overcomes it, and leads a happy, productive life depends mostly on personal factors (motivation, age of onset of the disability, general physical and mental condition, personal and institutional support). Many agencies, institutions, and foundations exist for the sole purpose of helping blind individuals. Privately funded agencies devoted to rehabilitation and other work with the blind are growing, as are the foundations devoted to supporting research work into the causes of blindness. This chapter outlines some of the services and programs available, but it is only a start. The most helpful hints, the most gratifying and useful personal guidance, reassurance, and support will probably come from a local agency and its counselors. Individuals living in large metropolitan areas can usually locate such agencies in the phone book or by referral from an eye doctor. The American Foundation for the Blind, Inc., 15 West 16th Street, New York, New York 10011, publishes a biennial *Directory of Agencies Serving the Visually Handicapped in the U.S.* If not available locally, it can be purchased from the foundation for $10. The American Foundation for the Blind also publishes the *International Guide to Aids and Appliances for Blind and Visually Impaired Persons* ($3). The catalog contains descriptions and ordering instructions for more than fifteen hundred devices ranging from Braille equipment to home, educational and occupational aids, games, sports, and personal equipment.

Optical Aids

Optical aids are magnifiers and telescopic devices that help the partially sighted. They are usually available through low-vision clinics or eye doctors who take a special interest in prescribing them. The devices themselves are described more fully in the section on optical aids in chapter 9. One's own eye doctor can usually make the referral to an appropriate clinic or office. If not, local agencies working with the visually disabled can do so.

Large-Print Books

An amazing number of publishers now produce titles in large print. The large-print editions of *The New York Times* and the *Readers Digest* are perhaps the most well known, but fiction, nonfiction, and textbooks also are available. The catalog *Large Type Books in Print* lists twenty-five hundred titles by various publishing houses. It can be obtained from the R. R. Bowker Co., Book Department, 1180 Avenue of the Americas, New York, New York 10036.

Local libraries usually stock a variety of large-print books and can get many more through the Library of Congress interlibrary loan network. The free booklet entitled "Volunteers Who Produce Books" lists agencies and groups who produce large-print copies, as well as many who transcribe into Braille or record the work. It also offers useful information on copyright, sources of information on books already available or in production, and suggestions on how to arrange for transcription of textbooks or foreign language books. For this pamphlet and a list of local libraries subscribing to the network, write to the Division for the Blind and Physically Handicapped, Library of Congress, Washington, D.C. 20542.

It is worth noting that most typewriter manufacturers produce large-type machines. In addition, many regular-type typewriters can be adapted to larger print. One should check with local typewriter sales or repair shops about the costs involved.

Braille, Mobility and Home Training, Rehabilitation, Financial Assistance

Braille is usually learned from teachers affiliated with local agencies serving the visually handicapped. Virtually every state and major metropolitan area in the United States has at least one agency listed by the American Foundation for the Blind. The handicapped person can get a direct referral from an eye doctor, a clinic, or a state rehabilitation division office. The state offices also advise on training and work opportunities for the visually handicapped.

Agencies for the blind also employ occupational therapists who help persons adapt well enough to the handicap to continue with home tasks, hobbies, and the like. Counselors train an individual to use a special walking cane for mobility and can recommend local guide-dog schools, where a blind person learns to work with the dog.

The federal Social Security Administration has assumed the

responsibility for financial assistance to the blind, and the local office should be consulted for details.

In Braille classes one usually learns of the wealth of books that can be secured from the Library of Congress and its participating network libraries. Non-textbook materials and Braille music scores are catalogued there. The Braille Book Bank, 422 S. Clinton Ave., Rochester, New York 14620, is a nonprofit organization that supplies Braille college-level textbooks, books on hobbies and general information, and cookbooks. They will send a catalog of their texts free of charge. Textbooks and writing equipment are available from the Instructional Materials Reference Center, American Printing House for the Blind, 1839 Frankfurt Ave., Louisville, Kentucky 40206.

Recorded Materials

The blind can take advantage of recorded books or cassettes, tapes, and special "talking book" 8-rpm recorders. The recorders, loaned free to the visually handicapped, can be obtained by applying to a regional branch library or by writing to the Division for the Blind and Physically Handicapped, Library of Congress, 1291 Taylor St., NW, Washington, D.C. 20542. Recorded books are available from the Library of Congress and its network of regional libraries. Completing a simple application form qualifies one for the recordings.

Recording for the Blind, Inc., is a nonprofit organization providing recorded educational books. Approximately forty thousand titles are available. A catalog can be obtained for $3 by writing to Recording for the Blind, Inc., 215 East 58th St., New York, New York, 10022. Choice Magazine Listening, 14 Maple Street, Port Washington, New York 11050, provides selections from periodicals every other month. Its free service consists of eight hours of selected articles, fiction, and poetry read by professionals onto the 8-rpm records that are used with the players provided by the Library of Congress.

New Developments and the Future

There is great promise for the future of the blind and near-blind now as new devices to promote mobility and aid in reading move from the drawing boards into practice. The most ambitious of them attempt to substitute for vision itself.

Laser canes are now being produced but are not yet in general use. They emit three laser light waves and by feedback in the form of auditory and tactile stimulation let the user know of drop-offs and obstacles that are ahead at body level, at head level, or at foot level. Bionic Instruments of Bala Cynwyd, Pa., produces these canes and will supply information on their cost and availability and on the training necessary for their use.

A sonar device is available through Telesensory Systems, Inc., of Palo Alto, Cal. Like a miniature radar set, this device receives and transmits auditory signals picked up from objects in its path. Size, consistency, and directionality are keyed to sound pitch, magnitude, and variable input for both ears. The device has been used with success in cases of infant and adult blindness. Called the Sonicguide, it consists of a transmitter of inaudible ultrasound waves and two receivers fitted into a lightweight spectacle frame. Reflected ultrasound waves are converted to audible signals and transmitted proportionately to each ear by tiny tubes that still allow ambient sounds to be heard. It detects obstacles above waist height and is used with a cane or guide dog.

The Pathsounder is a similar ultrasound device that is worn on the chest. It responds to objects above the waist up to about six feet away with audible sounds or tactile stimulation. An obstacle only thirty inches away will produce a higher-pitched warning sound or a vibratory stimulation of the neck. The Pathsounder is used with a long cane or guide dog.

The use of these three instruments requires special training by certified orientation and mobility specialists. Coordinating their purchase, maintenance, insurance and recycling and providing the training necessary to use them are the objectives of the nonprofit Mobility Foundation. This admirable young foundation is also attempting to bring the devices and training to the indigent blind. The task is enormous and the present level of funds inadequate, so private contributions are needed to fulfill their goals. Write to the Mobility Foundation at 745 Pine View Road, North Wales, Pa. 19454, for more information on its work and the devices mentioned.

SpeechPlus is a talking calculator developed by Telesensory

Systems. It costs several hundred dollars, but it can do simple and advanced calculations and operates from a sound keyboard arranged like a push-button phone.

Telesensory Systems also makes a reading machine that converts the photographic impulse received from a visual scanner used in one hand into a tactile stimulation under the opposite hand. It allows the average experienced user to read about forty or more words per minute. Called Optacon, the device is a small, portable battery-operated unit that transforms a letter under the camera scanner into a similar tactile array of elevated tiny rods under the user's index finger. It can be adapted to calculators, typewriters, cathode-ray tubes, and even fitted with magnifiers for extremely fine print. The Optacon costs several thousand dollars and requires special training at centers located throughout the world. Many public and private institutions are helping to distribute and finance Optacons. Information may be obtained from Telesensory Systems, Inc., 3408 Hillview Ave., P.O. Box 10099, Palo Alto, Ca. 94304.

Also available from Telesensory Systems is The Game Center, a collection of eight games for the visually disabled. The typewriter-sized machine talks to the player through headphones, advising of one's progress in tick-tack-toe, blackjack, skeet shoot, and other games played with hand-held control modules and a keyboard.

Demonstrated in December, 1977, was the exciting development of a true reading machine. Utilizing the hand scanning camera of the Optacon device coupled to a computer and speech synthesizer, this ingenious machine actually reads aloud to the blind person (figure 17–1). Telesensory Systems projects that a compact, marketable model of this prototype should be ready shortly.

Also under investigation now is adaptation of current television and computer technology for use by the blind. One such experiment involves transmitting a television image from a tiny portable camera, through a small portable computer that simplifies the picture, to an array of electrodes implanted surgically against the part of the brain that would normally receive visual stimulation. So far some patterns have been recognizable with the system. An-

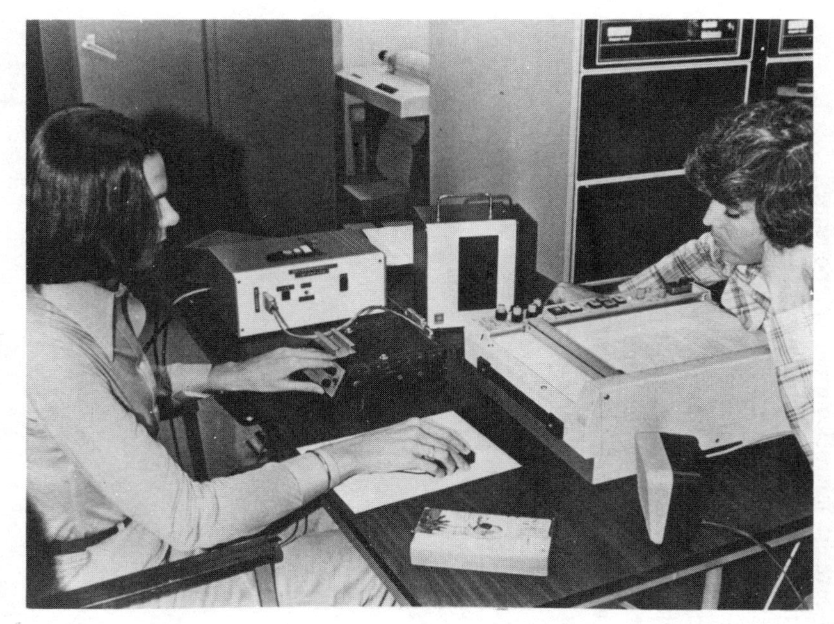

Fig. 17-1. Candace Linvill Berg hand-scans material with the Optacon camera while the computer at the Telesensory Systems, Inc. research facility reads the text being scanned to her in synthesized speech. Rob Savoie, research scientist in charge of the project, looks on. (Photograph courtesty of Telesensory Systems, Inc., Palo Alto, California.)

other concept converts the television picture received to an array of tactile stimulators, tiny electrodes that stimulate the skin in a pattern like a scoreboard. A tiny TV camera can be worn in a spectacle frame, and the tactile array worn against the back or abdomen. This system has been successfully adapted to a stationary microscope through which a blind person has been able to inspect and arrange microelectronic components.

Because the sophisticated technology and limited production of such devices will undoubtedly translate into rather high costs, large government and private contributions will probably be needed. The staggering economic and psychological costs of blindness and disability without rehabilitation, however, must be considered.

The future is hopeful for the blind. The many projects coming to fruition will lessen their disability and help them lead more active, enjoyable, and productive lives.

Index